Real-Life Math

Real-Life Math

everyday use of mathematical concepts

▲▼▲

EVAN M. GLAZER and JOHN W. McCONNELL

An Oryx Book

GREENWOOD PRESS
Westport, Connecticut • London

Library of Congress Cataloging-in-Publication Data

Glazer, Evan, 1971– .
Real-life math : everyday use of mathematical concepts / Evan M. Glazer and John W. McConnell.
p. cm.
Includes bibliographical references.
ISBN 0-313-31998-7 (alk. paper)
1. Mathematics—Popular works. I. McConnell, John W. II. Title.
QA93 .G45 2002
510—dc21 2001058635

British Library Cataloguing in Publication Data is available.

Library of Congress Catalog Card Number: 2001058635
ISBN: 0-313-31998-7

First published in 2002

Greenwood Press, 88 Post Road West, Westport, CT 06881
An imprint of Greenwood Publishing Group, Inc.
www.greenwood.com

Printed in the United States of America

The paper used in this book complies with the Permanent Paper Standard issued by the National Information Standards Organization (Z39.48-1984).

10 9 8 7 6 5 4 3 2 1

DEDICATED TO

the mathematics teachers at Glenbrook South High School
who were fearless in adapting real-world applications
to their lessons, and who generously shared
their best teaching ideas with us

Contents

▲▼▲

Introduction

▲▼▲

"When are we ever going to use this?"

This plaintive question from frustrated mathematics students is heard in schools around our country as they wrestle with pages of abstract mathematics and learn algorithms that appear to go nowhere. They study real numbers, but don't find any reason to believe that they are real. Thousands of American students still work from textbooks that limit applications to age problems and mixtures of nuts. Despite the call from the National Council of Teachers of Mathematics in the *Principles and Standards for School Mathematics* (2000) for meaningful learning through study of realistic applications, many students will find that the only modernization of content over their grandparents' math books is that jet planes have replaced the trains that used to travel at different rates between cities.

The twentieth century saw an explosion of applications of mathematics. It is now hard to find a field of study that does *not* use mathematical tools. Biologists use differential equations. Chemists use solid geometry to describe molecules. Set designers in theaters use trigonometry to determine the best lighting for a play. Historians determine authorship of obscure documents through statistical analysis of words. Governments, international corporations, and individual investors use mathematical rules to determine production, employment, and prices. Everybody uses computers. Unfortunately, even good students don't know how mathematics affects their lives. Few understand the power of compound interest. Few realize that the compound interest embedded in credit cards can bring adults to bankruptcy. Few know the mathematical implications of public policies that will affect their lives. Even fewer know how to make best decisions based on the probabilities of risk rather than blind gambles.

The secondary-school mathematics curriculum is faced with multiple challenges. What should students know and be able to do? Proficiency in some algorithms is important. Abstraction in mathematics—stripping concepts of all but their bare structures—is a feature that makes mathematics a powerful intellectual tool. But these are not sufficient. Much of the mathematics taught in grades 7 to 12 is there because it is important outside the math classroom. Foundation applications, like paths of projectiles, should not be stripped away, but rather should be used to motivate the arithmetic, algebraic, or geometric concepts. Further, students should have an opportunity to see a broad expanse of math applications so they can find links between their interests and aspirations and their mathematics coursework.

This book is an effort to promote real-world connections as they are applied in people's daily lives and careers. It is an account of the mathematical applications that we have learned and shared with people in our teaching careers. We hope this reference guide helps you enjoy and appreciate the use and application of mathematics in our culture and environment. We hope you will find some answers to the question, "When are we ever going to use this?"

audience

This book is intended to be a reference guide for anyone interested in understanding how some high school mathematics concepts are applied in nature and society. We hope that high school students, teachers, and librarians use these ideas to enhance their learning, teaching, and appreciation for mathematics. The mathematics described here cover concepts that are found in courses from prealgebra through introductory calculus. Each of the concepts is presented so that the reader can gain different levels of understanding due to the varying levels of mathematical complexity. A student or parent referencing the term *angle* will learn through descriptive text and diagrams that it is used for a variety of purposes in navigation and road construction. A student who has learned trigonometry may gain a deeper understanding as to *how* an engineer might use the mathematics to make predictions by viewing different formulas and calculations. Our intent is to make the content readable by all levels and ages of students, thereby hoping that they will recognize value in the applications of mathematics, regardless of their backgrounds.

purpose

This reference guide is an effort to provide exposure to mathematical applications, and should not be regarded as a primary tool for learning and instruction. Since we do not intend to teach mathematical concepts here, there are occasions in which mathematics is discussed without reference as to how an equation is formed or how it was solved. Instead, each concept is informally described so that primary emphasis can be placed on its applications. We do not intend for teachers to teach mathematics in the way it is presented here. Instead, the text should be used as a tool to enhance current instructional practices, or to spark student interest in math, or to create a classroom activity grounded in a particular application. Therefore, we feel that a more cohesive learning environment with these applications requires that the teacher and the learner examine the mathematical principles behind why *and* how a concept is applied.

content

The content in this reference guide is based on over forty mathematical concepts that are studied in different levels of high school mathematics. For example, linear functions are typically learned in algebra and are continually used beyond calculus. Each of the concepts is listed alphabetically and can be read independently. This format has been selected for pragmatic purposes, so that the applications can be used efficiently. Consequently, we occasionally synthesize concepts, such as referring to *slope* and *derivative* as rates, or cross-reference topics because some applications are based on related or multiple concepts.

The ideas presented in this book are not a comprehensive account of high school mathematics nor do they represent every possible application. We do not feel that every mathematical principle taught in a high school curriculum has a realistic application. We do feel there are situations where it is necessary to explore some mathematics that may not be applied. For example, the study of angles formed by parallel lines does not have many realistic applications, but the concepts can be used to introduce *similarity*, a topic with many useful applications. In addition, the concepts presented here do not introduce every application of high school mathematics. Our intent is to promote applications about mathematical concepts that are commonly studied in high school mathematics, even though there are additional interesting connections to other concepts that may not have as much emphasis in a school's curriculum. Furthermore, we simply cannot be aware of all of the applications that have realistic connections to the concepts we have listed. If you have any additional ideas, please share them with us by sending an email to <evanmglazer@yahoo.com>.

The depth of description of an application varies within each concept. Sometimes an application will be described in the form of a story, and other times it will be described in a few sentences to avoid redundancy with a similar analysis in another section. Sometimes we will just point in the direction of an important application. Sometimes we will provide a historical, rather than contemporary, application to show the genesis of a mathematical concept. The amount of mathematics described in each of the sections varies, depending on the context and complexity of the situation. We would much rather provide a flavor of how mathematics is used than go into detail for every application. In fact, many applications discussed in this book are based on simplified conditions, even though the real world often has unusual limitations, constraints, or peculiarities. For example, we neglect weather conditions when studying the motion of a baseball. Furthermore, we approximate the shapes of objects, such as assuming that the earth is a perfect sphere. Simplified situations are used in this reference guide in order to provide general principles in a concise manner so that the concepts can be understood by a high school student. World Wide Web references at the end of each section offer opportunities for further exploration of some of these applications. We offer such a listing here, giving Web references that provide a huge number of applications.

online sources for further exploration

The Math Forum
<http://www.mathforum.org/library/topics/applied/>

Contextual Teaching and Learning in Mathematics at the University of Georgia
<http://jwilson.coe.uga.edu/CTL/CTL/>

British Columbia Institute of Technology Mathematics Department Applications to Technology
<http://www.math.bcit.ca/examples/table.htm>

Micron's Math in the Workplace
<http://www.micron.com/content.jsp?path=/Education/Math+in+the+Workplace>

Mathematical Concepts

▲▼▲

ANGLE

Position, direction, precision, and optimization are some reasons why people use *angles* in their daily life. Street intersections are made at angles as close as possible to 90°, if not greater, so that visibility is easier when turning. It is beneficial for city planners to create additional turns so that there are larger turning angles for safer traffic. For example, if a car has to make a sharp 60° turn onto traffic, it would probably be more likely to get into an accident because the turn is difficult. If you find a nonperpendicular four-way intersection with a stoplight, it is likely to have a "No Turn on Red" sign for those drivers who would be at an obtuse angle. It would be easier for the driver if the road were constructed so that an additional intersection is added so the car can turn once at 150° and again at 90°.

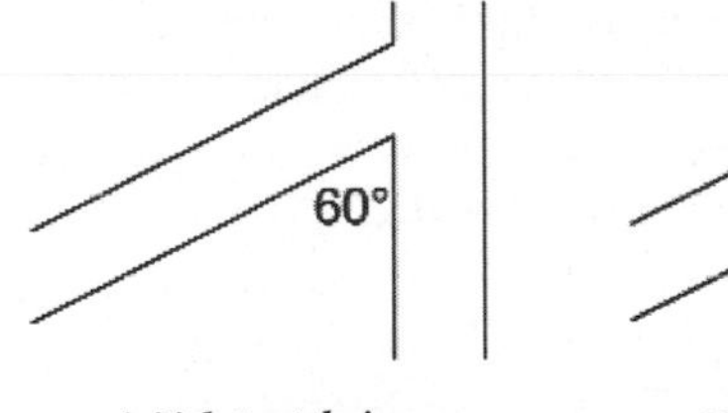

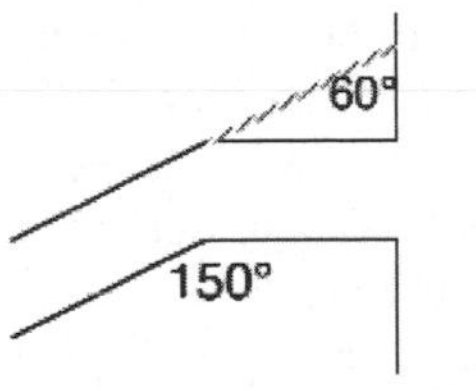

initial street design   *restructured street design*

Restructuring the angles in an intersection to make turning a vehicle easier.

The use of angles in the design of parking spaces affects how many cars can park in a lot. Most parking arrangements involve spaces that are perpendicular or slightly angled to the curb. An advantage to using obtuse-angled spaces is that it is easier to turn a car at an obtuse angle than at a right angle, so there may be less accidents in a lot with angled spaces. An advantage to using right-angled spaces is the opportunity to fit more cars in the parking lot.

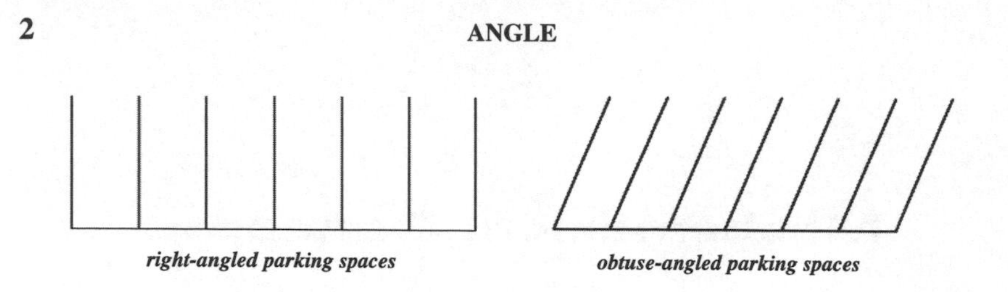

Parking-space arrangements in parking lots.

The amount of space, s, saved by using right-angled spaces is $s = -l \cos \alpha$ for each row in the parking lot, where l is the length of the space and α is the angle of the turn into the space. When the shape of a space is transformed from a rectangle (right-angled) to a parallelogram (obtuse-angled), the extra horizontal distance needed in a parking-lot row will be the amount of space that the last car displaced from its previous perpendicular arrangement. In the obtuse-angled situation, the length of the parking space is the hypotenuse of a right triangle formed with the curb. The cosine of the angle between the curb and the parking lines, $\cos \theta$, is the ratio of the horizontal curb space, s, to the length of the parking space, l. In an equation, this is written as $\cos \theta = \frac{s}{l}$.

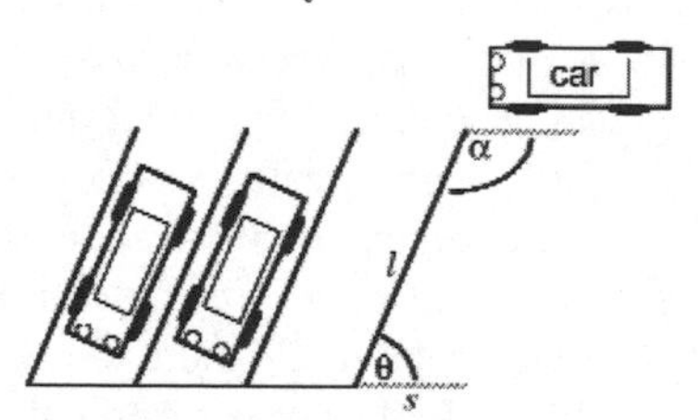

Variables that affect the extra horizontal space, s, that is needed in a parking lot with angled spaces.

Multiplying both sides of the equation by l will change it to $s = l \cos \theta$. The angle against the curb and the car's turning angle are supplementary, because the curb and car's path are parallel. The interior angles on the same side of the transversal (the parking lines) are supplementary, so $\cos \theta = -\cos \alpha$. Substituting this result into $s = l \cos \theta$ generates the equation, $s = -l \cos \alpha$.

If the parking lines were at a 60° angle with the curb, the turning angle would be 120°. Suppose the dimensions of a parking space are 8 feet by 20 feet. If the lot is transformed from right-angled spaces to oblique-angled spaces, each row would lose $s = -20 \cos 120° = 10$ feet, which is equivalent to a little more than one space!

An overhead view of a car making an obtuse-angled turn of α degrees into a parking spot that is angled θ degrees with the curb.

Airplane pilots, military-orienteering specialists, and ship-navigation crews are responsible for using angles to move efficiently towards a destination. After accounting for wind and current speed, navigation teams will determine an angle to direct their course of movement. For example, suppose a ship is 3 miles from shore and is docking at a port that is 6 miles away, with minimal current affecting the path of the boat. The captain will request the boat to be moved 60° West of North, or 30° North of West. This direction is equivalent to the angle that is formed between the path of the boat and the northern or westward direction. The captain can also simply ask to move the boat 60°, because it is assumed that navigation direction is counterclockwise from the North position.

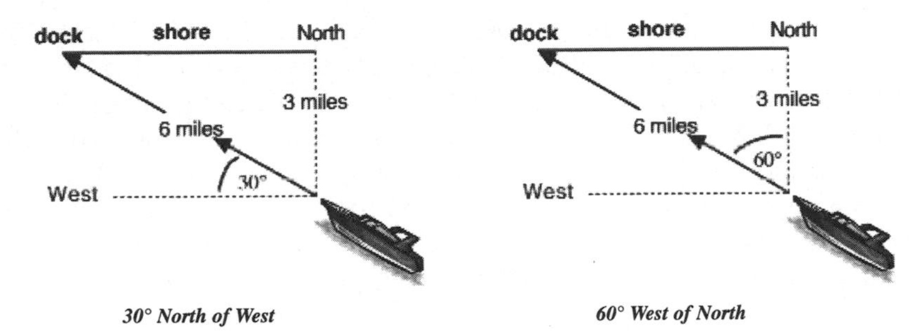

Alternate methods of denoting direction when a boat is 3 miles from shore and 6 miles from its destination.

Notice that the distances from the port are represented in a 30°–60°–90° triangle, which will not always happen. The angle of navigation, β, that will be East or West of North can be determined by finding $\arccos(\frac{s}{d})$, where s is the distance from shore and d is the distance to the final destination. Notice that the navigation angle will be negative, or East of North, if the destination is situated to the East of the ship's location.

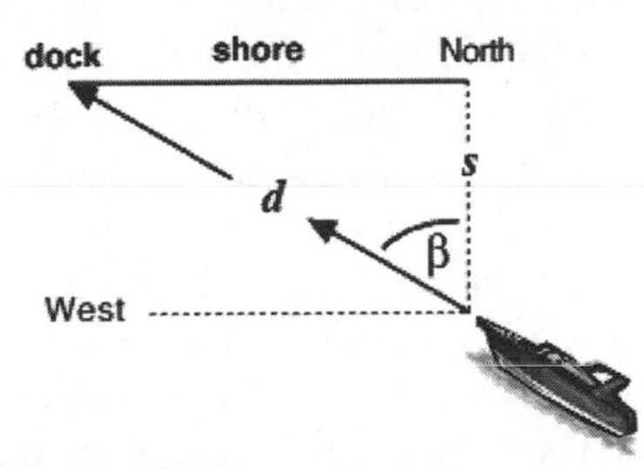

The navigation angle, β, of a boat based on its distance from shore, s, *and the distance from its final destination,* d, *is equal to* $\arccos(\frac{s}{d})$.

Sailboats cannot steer directly into the wind, because they would be pushed backwards. In order to sail against the wind, sailors need to tilt their boat at an angle, ideally 45°, so that the wind catches the sail. If the boat pushes off course, it will need to change direction again so that it moves perpendicular to its path in the opposite direction. Sailors call this *tacking*. This action ensures that the boat continues to maintain a 45° angle with the wind. This zigzag pattern enables the boat to reach its finishing point while constantly sailing into the wind.

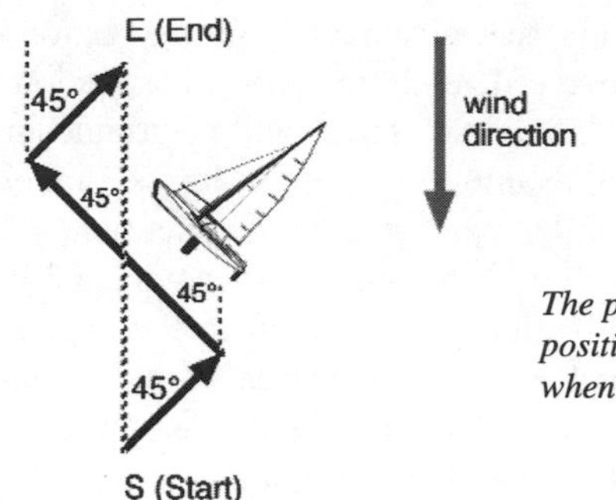

The path of a boat from its starting position, S, to its ending position, E, when it is sailing against a headwind.

Angles are useful for reflecting light rays or objects off of flat objects. The angle by which an object, such as a ball, approaches a wall is equal to the angle by which the object bounces off the wall. This is true because a ball's reflection off a wall will be the same distance away from the wall as if it had gone in a straight line. In essence, reflections preserve congruence. By the transitive property, the angle of the ball coming into the wall will equal the angle of the ball leaving the wall, as shown below. In billiards or miniature golf, a player can use this principle when aiming for a hole by simply aiming for the hole's reflection.

The angle of a ball approaching a wall will equal the angle of the ball leaving the wall, assuming there is no spin on the ball.

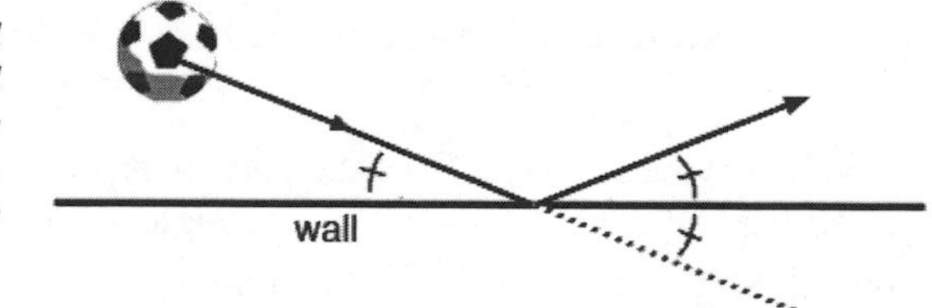

Athletes who try to throw or hit balls certain distances, such as baseballs, basketballs, footballs, and golf balls, use angles strategically. If they want to hit a ball short and high, they will use an angle close to 90°. In order to hit a low-flying line drive, they will use an angle close to 0°. The horizontal distance in meters, x, of an object can be determined by the product of its initial velocity in meters per second, v_0, the time in seconds, t, that the ball is in the air, and the cosine of the angle, α, it is released or hit. Since the earth's gravitational force pulls a ball towards the surface, the vertical distance in meters, y, also needs to be considered in order to determine the ideal angle at which to release or hit a ball. The two equations describing the path of the ball in both directions are represented as

$$x = v_0 t \cos \alpha$$
$$y = v_0 t \sin \alpha - 4.9t^2.$$

The ball will be on the ground when y is 0. Solving the second equation for the time t that will provide this value gives $t = 0$ or $t = \frac{1}{4.9} v_0 \sin \alpha$. The latter solution gives the time the ball will be in the air. Substituting in the equation for x

yields $\frac{1}{4.9}v_0^2 \sin a \cos a$. Using trigonometric identities gives $x = \frac{v_0^2 \sin(2\alpha)}{9.8}$. Suppose a golfer hits a tee shot, and that his or her club hits the ball at $v_0 = 70$ meters/second. The graph of horizontal distances x as a function of the angle α shows that the angle that will give the golfer the best distance is 45° ($\pi/4$ radians). Frogs know this angle: push-off angle for a frog hop has been measured to be close to 45°.

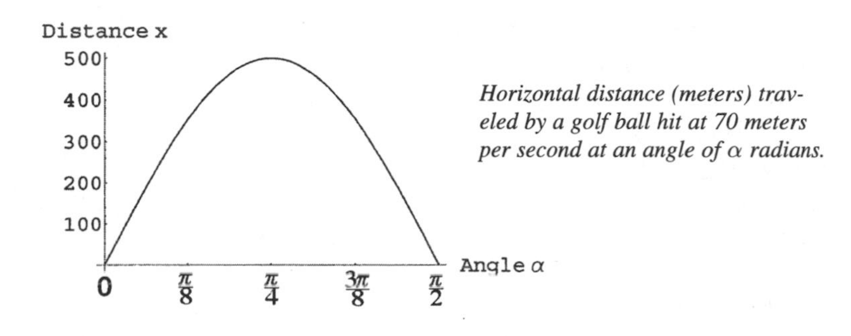

Horizontal distance (meters) traveled by a golf ball hit at 70 meters per second at an angle of α radians.

When a golfer tees off or a football kicker aims for a long field goal, he or she should strike the ball at a 45° angle in order to obtain maximum distance. A baseball player, on the other hand, needs to alter this thinking slightly, because he hits a ball about 1 meter off of the ground. This makes the horizontal-distance equation more complicated:

$$x = v_0 \cos\alpha\left(\frac{v_0 \sin\alpha\sqrt{v_0^2 \sin^2\alpha - 19.6(y-1)}}{9.8}\right).$$

When the ball hits the ground ($y = 0$), the graph of this function shows that a ball reaching the bat at 85 miles per hour, or 38 meters per second, will attain a maximum horizontal distance when the ball leaves the bat at about a 44.8° angle, very close to the angle if the ball were hit from the ground.

The refraction of light is dependent on the angle in which light enters the object and the material it passes through. *Snell's law* states that $n_1 \bullet \sin\Theta_1 = n_2 \bullet \sin\Theta_2$, where n is the index of fraction (the ratio of the speed of light in air to the speed of light in that material) and Θ is the angle of incidence. As light passes through an object, such as a glass of water, it will bend, giving it a distorted view if you look through the glass. Higher values of n allow the light to bend more, since Θ_2 decreases as n_2 increases.

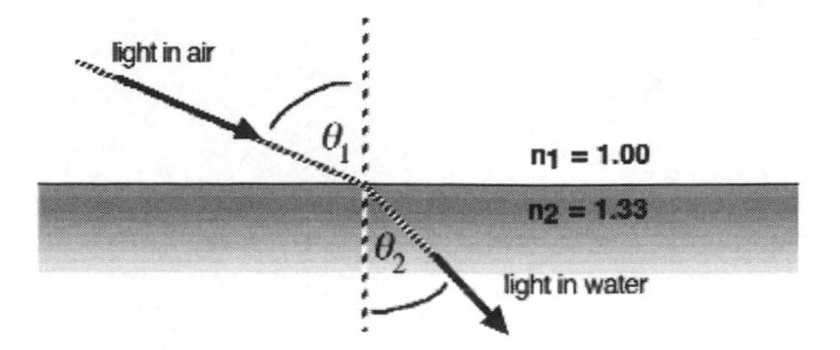

The angle of light rays will change after hitting a different surface, such as water. Snell's law can be used to determine the angle of refraction, or the angle in which light bends as it passes through a new surface.

Gems such as diamonds have a high index of refraction, allowing them to trap light and reflect it internally, which consequently makes them sparkle.

online sources for further exploration

The best angle to view a baseball game:
<http://forum.swarthmore.edu/pow/solutio65.html>

Diamond design:
<http://www.gemology.ru/cut/english/tolkow/_tolk1.htm>

Finding your way with map and compass:
<http://mac.usgs.gov/mac/isb/pubs/factsheets/fs03501.html>

The mathematics of rainbows:
<http://www.geom.umn.edu/education/calc-init/rainbow/>

Navigation problems:
<http://jwilson.coe.uga.edu/emt725/Bearings/Bearings.html>

Photography angles:
<http://www.a1.nl/phomepag/markerink/shifcalc.htm>
<http://www.a1.nl/phomepag/markerink/tiltcalc.htm>

Projectile motion simulations:
<http://library.thinkquest.org/2779/Balloon.html>
<http://www.explorescience.com/activities/Activity_page.cfm?ActivityID=19>
<http://www.phys.virginia.edu/classes/109N/more_stuff/Applets/ProjectileMotion/jarapplet.html>

River crossing–swimming angles:
<http://www.emsl.pnl.gov:2080/docs/mathexpl/swimwalk.html>

Robot navigation angle:
<http://www.ezcomm.com/~cyliax/Articles/RobNav/robnav.html>

Sailing strategies:
<http://www.orfe.princeton.edu/~rvdb/sail/sail.html>

Snell's law:
<http://www.physics.nwu.edu/ugrad/vpl/optics/snell.html>
<http://www.yorku.ca/eye/snell.htm>
<http://www.glenbrook.k12.il.us/gbssci/phys/Class/refrn/u14l2a.html>
<http://buphy.bu.edu/py106/notes/Refraction.html>

Throwing a boomerang:
<http://www.concentric.net/~davisks/throwing/>
<http://www.bumerang-sport.de/throwing/throw.htm>

ASYMPTOTE

An *asymptote* is an imaginary line or curve that a function approaches as its independent variable approaches infinity or an undefined value. A *vertical asymptote* of $x = c$ exists on a function $f(x)$ if at a point of discontinuity, $x = c$, the limit of $f(x)$ as x approaches c equals positive or negative infinity. A *horizontal asymptote* of $y = k$ exists on a function $f(x)$ if the limit of $f(x)$ as x approaches positive or negative infinity equals k. For example, the function $f(x) = \frac{x-2}{x+3}$ has a vertical asymptote at $x = -3$ because $\lim_{x \to -3} \frac{x-2}{x+3} = \pm\infty$ and a horizontal asymptote at $y = 1$ because $\lim_{x \to \pm\infty} \frac{x-2}{x+3} = 1$.

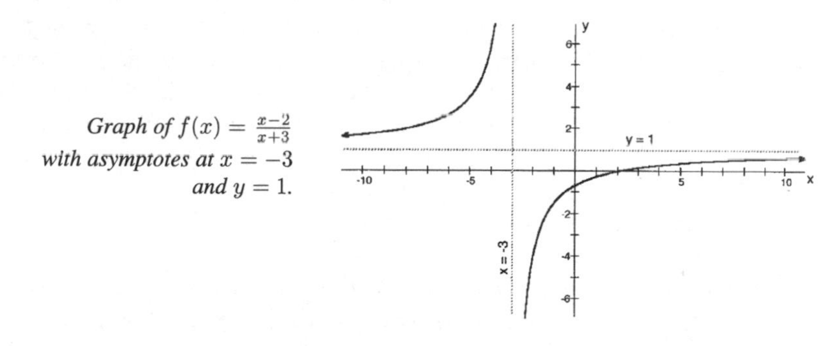

Graph of $f(x) = \frac{x-2}{x+3}$ *with asymptotes at* $x = -3$ *and* $y = 1$.

In the real world, horizontal asymptotes typically represent a leveling-off effect, such as the radioactive decay of a particle diminishing until it is almost gone. (See ***Exponential Decay***.) If the dependent variable y is the amount of the particle, then in this case there would be a horizontal asymptote of $y = 0$ on the graph, because the amount of the particle approaches zero. Most substances that have a decaying effect, such as the amount of power supply in a battery, will have an asymptote of $y = 0$ on a graph that describes its amount as a function of time.

The cooling of hot liquids in a mug, such as coffee, illustrates asymptotic behavior because the liquid gradually approaches room temperature after sitting awhile in the cup. The warming of liquids, such as ice sitting in a cup, demonstrates a similar phenomenon, except that the temperature graph rises towards the

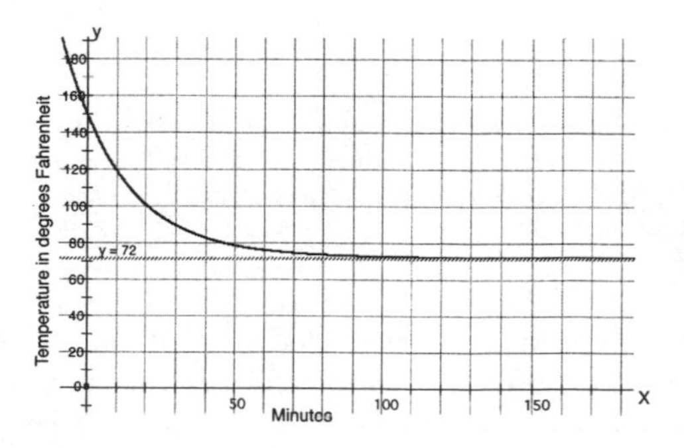

A graph of the temperature, in degrees Fahrenheit, of coffee as a function of the number of minutes it sits in a closed cup. The temperature of the coffee levels off near room temperature after an hour and a half.

asymptote. In both cases, the asymptote would represent the room temperature, because the liquid either warms or cools to that temperature after it is left out for awhile.

Scientific barriers based on speed are asymptotic until technological advances overcome a barrier. For example, airplanes could not pass the sound barrier, called Mach 1, until 1947. (See ***Ratio***.) Before that time, airplanes progressively became faster and faster, approaching the speed of sound but unable to surpass it, because they were not built to handle the shock waves produced at such speeds. However, once the barrier was broken, scientists and engineers were given data that helped them develop airplanes that could maintain their structural integrity under the stressful conditions associated with travel at those speeds. Today, particle physicists are challenging the speed of light by accelerating particles in large circular chambers. As testing and experimentation progresses over time, the detected speeds of particles have been gradually approaching the barrier of 3×10^8 meters per second. Scientists argue whether it will be possible to move at speeds faster than light, and if so, what type of consequence will occur. Many science-fiction stories portray ships disappearing when they travel faster than the speed of light, because light is not fast enough to show an image of the ship to an observer.

Terminal velocity is the limiting speed of an object due to wind resistance when it is in free-fall. For example, a skydiver will jump out of an airplane and be pulled towards the earth at an acceleration of 9.8 meters per second squared. This means that the velocity of the person falling will gradually increase until it reaches terminal velocity. The equation $v = 9.8t$ describes the velocity, v, in meters per second of a person falling out of the plane after t seconds. After 1 second, the skydiver is falling at a rate of 9.8 meters per second, and after 2 seconds, the person's velocity has increased to 19.6 meters per second. However, if the skydiver lies flat during free-fall, the wind resistance will inhibit the falling rate so that the body does not exceed 50 meters per second. Consequently, $y = 50$ becomes the horizontal asymptote on the velocity versus time graph. This information is helpful for the skydiver to determine how much time can be spent in the air for skydiving acrobatics and at what point the parachute should be opened for safe landing.

A person jumping from an airplane will reach a terminal velocity at which he cannot fall any faster due to wind resistance.

Vertical asymptotes typically appear in applications that deal with improbable events, costs, or quantities. For example, the cost to extract petroleum from the Earth is dependent on its depth. Oil that is deeper underground will typically be more expensive to remove, because it is more difficult to create deeper tunnels. In such situations, workers take an increased risk of the tunnel caving in, as well as having to deal with the added distance covered by equipment to extract dirt and rocks. This means that tunneling down 11 to 20 feet may be twice as difficult than tunneling the first 10 feet; and tunneling down 21 to 30 feet may be three times as difficult than tunneling down 11 to 20 feet, and so on. Consequently, a vertical asymptote will exist near the deepest level on a graph, indicating that it would be nearly impossible to dig at certain depths. Geologists would find this information useful, thus being able to recommend the appropriate digging depths that would be safe and economically beneficial to the government and local business.

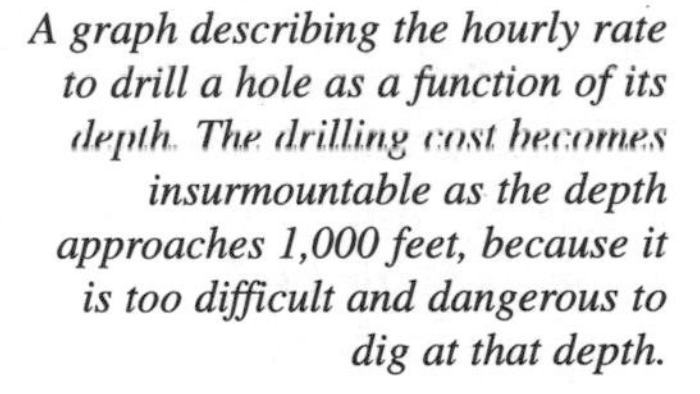

A graph describing the hourly rate to drill a hole as a function of its depth. The drilling cost becomes insurmountable as the depth approaches 1,000 feet, because it is too difficult and dangerous to dig at that depth.

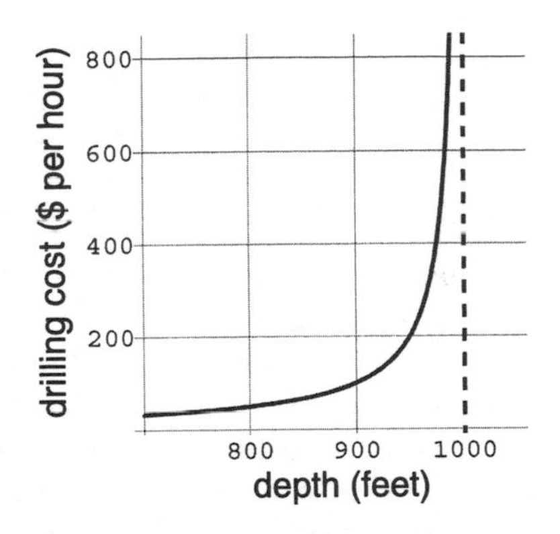

Vertical asymptotes also sometimes symbolize quantities that are nonexistent. For instance, if a preschool wants to build an enclosed playground for its students, it would need to build a fence. The builders would consider the best way to lay out their available fencing so that the students have a large amount of play space. A function to describe the dimensions of a rectangular play space are $w = \frac{200}{l}$, where w is the unknown width of the field, l is the unknown length of the field, and 200 square feet is the desired area of the play space. In this case, $l = 0$ is a vertical asymptote, because not only is it impossible to divide by zero, but it is impossible to have a rectangular play space that does not have any length!

online sources for additional exploration

The basics of cooling food
<http://www.hi-tm.com/Documents/Basic-cool.html>

Investigate the behavior of northwestern crows
<http://illuminations.nctm.org/imath/912/Whelk/index.html>

Modeling of disease and disease progression
<http://www.phm.auckland.ac.nz/Staff/NHolford/Mss/Disprog/modelling_disease-progression.htm>

Scaling the Internet Web servers
<http://www.cisco.com/warp/public/cc/pd/cxsr/400/tech/scale_wp.htm>

The terminal velocity of coffee filters
<http://aci.mta.ca/TheUmbrella/Physics/P3401/Investigations/VterminalDDB.html>

Time travel?
<http://members.aol.com/JLandGDC/numin/1999/oct99.htm>

The twisted pendulum experiment
<http://www.carolina.com/coachlab/math.asp>

▲▼▲

CARTESIAN COORDINATES

Coordinates are useful to determine relative position and distances. For example, *pixels* (dots of light) on a computer are identified by their horizontal and vertical components, where (0,0) is at the corner of the screen. The coordinates of the pixels are useful for animations that require starting and ending points for each vertex in a diagram. Given this information, the computer will predict intermittent coordinates of the vertices to help render the animation, without having to input the coordinates for every second on the screen.

Coordinates are also useful in computer programming to plot points on the screen or define regions on a blueprint or graphic. For example, an *image map* is a graphic that links certain portions of a Web page to different pages on a Website. Image maps are used to enhance the colors on a screen or to provide a larger region to click a list of items. The image map will probably look like a series of buttons that are defined by geometric regions, such as rectangles or circles. When the cursor is moved to a coordinate within a defined region on the image map, it will move to a new page once the mouse is clicked. Suppose a rectangular region is defined so that its upper-left coordinate is (12,35) and lower-right coordinate is (40,70), as shown in the illustration below. This will create a *hot spot* region

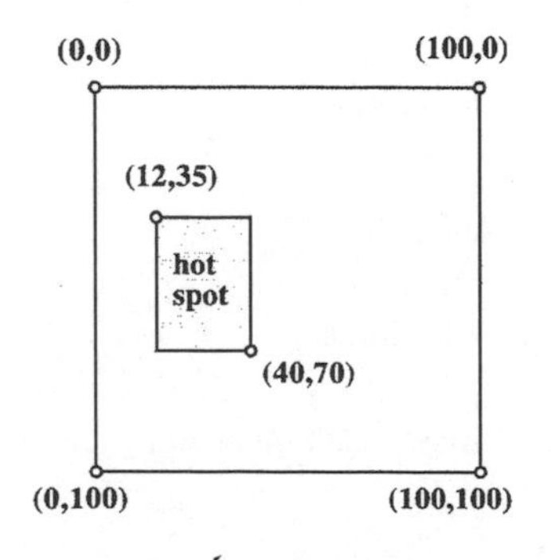

An image map uses coordinates to define a rectangular "hot spot" by noting the opposite corner coordinates in an image map.

with dimensions of 28 pixels by 35 pixels that will link to a new page if the cursor is clicked at a location on the image map between 12 and 40 pixels and between 35 and 70 pixels. If the cursor is not in this region, then it will not link to that page. Notice that the coordinate system on the image map is defined differently from the standard rectangular system. Since only positive values are used, this coordinate system uses the opposite of the negative y-coordinates that are represented in the fourth quadrant of a Cartesian coordinate system.

Desirable locations for fire stations are places where trucks would have equal access to the entire town. Ideally, they should be situated so that the longest drive to the edge of town is the same in all directions. A coordinate grid could be superimposed on a city map, assigning coordinates to each of the intersections. The distance formula, $d = \sqrt{(x_1 - x_2)^2 + (y_1 - y_2)^2}$, could then be used to determine relative distances, d, of each street based on the coordinates of its endpoints, (x_1, y_1) and (x_2, y_2), so that the best possible intersection for the fire station could be selected.

On a world map, cities and landmarks are assigned a position according to how far away they are from the equator (0°latitude) and from the prime meridian in Greenwich, England (0° longitude). For example, Chicago is near 41° N 87° W, which means that it is $\frac{41}{90}$ in the northern hemisphere and $\frac{87}{180}$ in the western hemisphere.

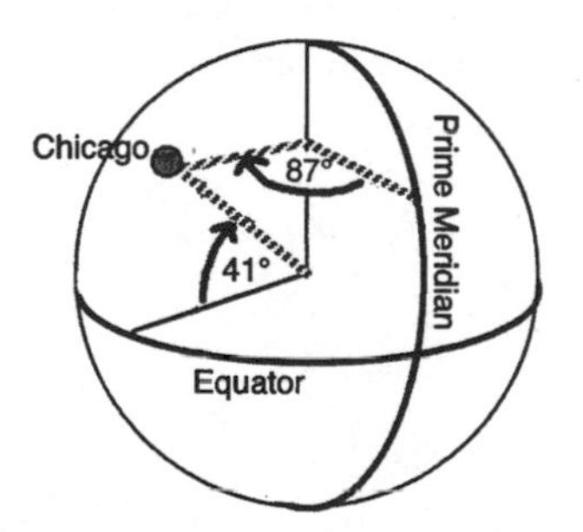

The position of Chicago on the earth according to its latitude and longitudinal positions, which are the same as the central angles from the center of the earth.

A flight from Chicago to Los Angeles would angle the plane 7° south of west and expect to travel 31° westward on its journey, because Los Angeles is near the position 34° N 118° W. The distance d traveled between any two cities on the globe can be determined by the equation

$$d = 3963 \arccos[\sin(\text{latitude}_1)\sin(\text{latitude}_2) + \cos(\text{latitude}_1)\cos(\text{latitude}_2)\cos(\text{longitude}_2 - \text{longitude}_1)],$$

where the position in a spherical coordinate system of two cities are $(\text{latitude}_1, \text{longitude}_1)$ and $(\text{latitude}_2, \text{longitude}_2)$ in radians. There are 2π radians in 360°, so each coordinate should be multiplied by $\frac{2\pi}{360}$ to convert to radians. In this case, the Chicago coordinate would convert from (41,87) to approximately (0.7156,1.5184), and the Los Angeles coordinate would convert from

(34,118) to approximately (0.5934,2.0595). Using this formula, the distance between Chicago and Los Angeles is

$$d = 3963 \arccos[\sin(0.7156)\sin(0.5934)+ \cos(0.7156)\cos(0.5934)\cos(2.0595-1.5184)],$$

which is approximately 1,758 miles.

online sources for further exploration

Celestial coordinates
<http://www.lhs.berkeley.edu/SII/SII-FindPlanets/SII-FindThatComet/coordinates.html>

Creating an image map
<http://www.personal.psu.edu/users/k/x/kxs156/tuthow.htm>
<http://www.ils.unc.edu/utils/imagemap-tutorial.html>

Georeferencing and digital images
<http://magic.lib.uconn.edu/help/aerialphotos/GeoreferncingAndDigitalImages.html>

The satellite times
<http://celestrak.com/columns/v03n02/>

Spherical coordinates and the GPS
<http://www.math.montana.edu/frankw/ccp/cases/Global-Positioning/spherical-coordinates/learn.htm>

Stereograms
<http://library.thinkquest.org/2647/misc/stertech.htm>

▲▼▲

CIRCLES

Circles are used in many real-world applications. All manholes are round so that their covers never slip through the pipes from the ground to the sewers. Any way you turn the cover it is impossible to force it through the hole, since the distance from the center of the circle is always the same. Since polygons do not hold this property, a circle is very useful for this purpose.

Circular wheels allow the opportunity for constant and smooth motion when riding a bicycle or automobile. If the circle had edges or vertices the ride would become very bumpy, because the distance from the center of the wheel to its perimeter would no longer be constant. In addition, a car will travel the distance its wheels rotate, because the friction between the wheel and pavement cause the car to move. For every revolution the tires make, the car will travel the length of the circumference of them. If a wheel has a diameter of 32 inches, then its circumference, or distance around, is 32π inches ≈ 100.5 inches.

In addition to distance traveled by an automobile, the circumference of circles is used in several applications. A trundle wheel is a device used to measure distances that are too long for a tape measure. A marking is placed on the wheel so that it clicks for one complete revolution. A trundle wheel can be made in any size, although it is convenient to make one with a diameter of 31.8 cm, because then its circumference will be 1 meter (circumference is the product of π and the diameter of the circle). Therefore, as you push the trundle wheel, every click that is recorded on the odometer means that the wheel has gone around once and you have traveled 1 meter.

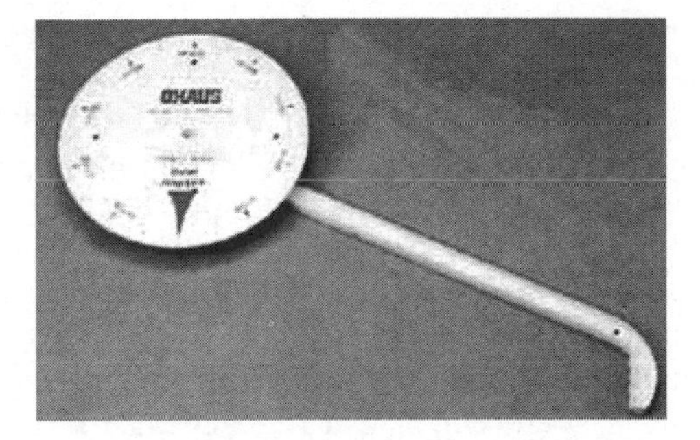

A trundle wheel can be used to measure long distances by multiplying the number of its revolutions by its circumference.

Another useful application of circumference is to determine the age of certain trees. The girth, or thickness, of trees increases as they grow older. A fallen tree often shows a large group of concentric rings, where each ring represents a year of its life. Since the tree gets thicker during its lifetime, the number of rings is proportional to its circumference. Therefore, a functional relationship can be created to estimate the age of a tree based on its circumference. This means that a measurement of the circumference of a tree can give an indication of its age without having to chop it down and count its rings.

The age of a tree is related to the number of its rings and its circumference.

The area of a circle is useful to determine the price of circular foods that have the same height. For example, pizzas are often advertised according to their diameter. A pizza with a diameter of 12 inches might sell for $10, and a pizza with a diameter of 16 inches for $16. Is that a reasonable deal? Since the amount of pizza is related to its area, it would be more beneficial if the consumer were told the unit cost of the pizza per square inch. Instead, consumers may develop a misconception and think that the 16-inch pizza should be 16/12, or 4/3, as much as the 12-inch pizza.

In the 12-inch pizza, the radius is 6 inches. So the area of the pizza is $\pi(6)^2 \approx 113.1$ square inches. At a sale price of $10, the consumer is paying about 8.8 cents per square inch of pizza. In the 16-inch pizza, the radius is 8 inches. So the area of the pizza is $\pi(8)^2 \approx 201.1$ square inches. At a sale price of $16, the consumer is paying about 8.0 cents per square inch of pizza. At first glance, one might think the 12-inch pizza is a better buy, but actually it is the other way around. Since volume purchases usually have a cheaper unit price, these prices seem pretty reasonable. Is this true about the prices at your favorite pizza shop?

The area of a circle is helpful to farmers in determining the amount of space that a sprinkling system will cover. As a sprinkler rotates, it will spray water in a circular pattern, or in a sector of a circle if it is restricted in a certain way. The distance the water reaches, or the radius of the circle, is sufficient information for the farmer to determine how much space will be covered by the water and how many sprinklers are needed to water the crops. Crops are often created in rectangular grids to make harvesting easier, but watering in a rectangular pattern is often less efficient than in a circular pattern. Therefore, the challenge in watering crops is to determine how many circles can be packed into the rectangle region. The trick for the farmer is to automate the sprinklers so that they provide just the right amount of water to the crops to optimize production and minimize expense.

A circle is a figure that has an optimal area based on its perimeter. Based on a given perimeter, there is not another shape that has an area greater than a circle. Similarly, based on a given area, there is not another shape that has a smaller perimeter than a circle. In essence, this information indicates that a great way to make use of materials and space is to form circles. Think about all the objects made of raw materials that are shaped into circles, such as plates, cups, pots, compact discs, and digital video discs. All of these objects are designed to hold substances or information that take up space in a resourceful way. Parts of circles

A keystone at the center of an arc above a doorway maintains its structure and support.

can also be used for aesthetic design purposes, such as the arches seen over some doorways. Roman engineers mastered the use of the circular arch in buildings, bridges, and aqueducts. A keystone, the stone placed at the top of the arch, is the essential component that keeps the structure of the arch together. Without a keystone, the arch may crumble if it is not cemented properly.

All materials are not designed to include circles, however, because a circle does not necessarily serve all functions. For example, a book is shaped like a rectangular prism instead of a cylinder, because it may be easier to store on a shelf and retrieved easily with its visible binding.

Circular, or angular, motion has several useful applications. It affects the linear speed and performance of many objects. For example, circular disks spin in an automobile engine to move its timing belts. The size of the disks can vary, allowing the engine to distribute its power in different ways. In order to move a belt, larger wheels do not need to spin as fast as smaller wheels, because they cover a greater distance in a smaller amount of time. (See ***Variation***.)

Another way to think about the connection between angular and linear speed is to envision the motion of an ice skater. The spinning rate of the skater will change with the movement of the radius of his or her arms from the body. To move faster, the skater will pull his or her arms in towards the body; conversely, to spin more slowly, the skater will gradually pull his or her arms away from the body. As an equation, the linear speed, s, is the product of the radius, r, and angular speed, ω, written as $s = r\omega$. Suppose the skater has a constant linear speed of 500 cm/sec. If his or her arm radius is 100 cm, then the skater will be spinning at a rate of 5 radians/sec, or less than 1 revolution in a second. If he or she pulls the arms in so that they are 25 cm from the body, then the skater's angular speed picks up to 20 radians/sec, about 3 1/2 revolutions in 1 second.

The skater slows her camel spin by extending her arms.

If the angular speed is held constant, then an object can have different linear velocities depending on its position on the circular object. For example, a spinning object on a playground or at an amusement park, such as a merry-go-round, typically has a constant angular speed. Therefore linear velocity increases as the radius increases. This means that you would feel like you were moving faster if you stood further away from the center. If you like rides that make you feel dizzy, then make sure you stand near the outside of a circular wheel when it is in motion.

online sources for further exploration

The arch in architecture
<http://www.ba.brantacan.co.uk/architecture.htm>

Circular motion
<http://www.glenbrook.k12.il.us/gbssci/phys/Class/circles/u6l1e.html>
<http://www.sd83.bc.ca/stu/9906/agal_3b.html>

Make your own trundle wheel
<http://www.geocities.com/thesciencefiles/trundle/wheel.html>

Pizza prices
<http://www.ecst.csuchico.edu/~pizza/pizzaweb.html>
<http://www.mrpizzaman.com/pizza/index.html>
<http://www.panola.com/biz/pizzahut/create.htm>

Tree rings
<http://www.geo.arizona.edu/K-12/regression/>
<http://web.utk.edu/~grissino/>
<http://www.ngdc.noaa.gov/paleo/treering.html>

CIRCUMFERENCE. See CIRCLES

▲▼▲

COMPLEX NUMBERS

Complex numbers are numbers expressed in the form $a + bi$, where a is the real number component and b is the imaginary number component. The number i is the square root of negative 1: $i = \sqrt{-1}$. Numbers in the physical world are often represented by their real number component, such as in measurement, money, and time. For example, a mile is a unit of measurement that is equivalent to 5,280 feet. As a complex number, this measurement would be $5{,}280 + 0i$ feet. However, the expression in complex form does not produce any additional meaning if the imaginary number component is equal to zero. Therefore, complex numbers are useful when the imaginary number component is nonzero.

There are several instances in which imaginary numbers are important in the physical world. For example, some circuits have unexpected changes of voltage when introduced to current and resistors that have imaginary number components. The amount of voltage in a circuit is determined by the product of its current and resistance. Without an imaginary number component in both current and resistance, the voltage reading will remain unaffected. For example, suppose the current is reading 3 + $2i$ amps on a circuit with 20 ohms of resistance. The net voltage would be (3+$2i$)(20) = 60 + $40i$ volts. In this case, the voltmeter would show a reading of 60 volts, because the $40i$ volts are imaginary. However, if the resistance was 20 + $4i$ ohms, then the net voltage would be (3 + $2i$)(20 + $4i$) = 60 + $12i$ + $40i$ + $8i^2$. Since $i^2 = -1$, this expression simplifies to 52 + $52i$. That means that the introduction of an imaginary number component in the resistance of the circuit would result in a voltage drop of 8 volts!

Electromagnetic fields also rely on complex numbers, because there are two different components in the measurement of their strength, one representing the intensity of the electric field, and the other the intensity of the magnetic field. Similar to the electric circuit example, an electromagnetic field can have sudden variations in its strength if both components contain imaginary components.

Complex numbers also indirectly have applications in business. The profit of the sales of a product can be modeled by a quadratic function. The company will start with initial expenses and rely on the sales of their product to transfer out of debt. Using the quadratic formula, the business can predict the amount of sales that will be needed to financially break even and ultimately start making a profit. If complex zeroes arise after applying the formula, then the company will never break even! On a graph in the real plane, the profit function would represent a parabola in the fourth quadrant that never touches the horizontal axis that describes the number of products sold. This means that the business will have to reevaluate their sales options and generate alternative means for producing a profit.

To generalize this case, any quadratic model that produces complex solutions from an equation will likely indicate that something is not possible. For example, in the business-sales setting, the company may want to test when the profit will equal one hundred thousand dollars. When solving the equation, the quadratic equation could ultimately be applied, and the existence of imaginary components in the solution would verify that this would not be possible. The same argument could be applied to determine if the world's strongest man could throw a shot put 50 feet in the air. If a person can estimate the throwing height h_0 and the time t the ball is in the air, then the quadratic function $h = 0.5gt^2 + v_0t + h_0$ can be applied to determine the initial velocity v_0 and whether the ball will reach a height h of 50 feet. (Note that the gravitational constant g on earth is equal to –9.8 meters per second^2, or –32 feet per second^2.)

online sources for further exploration

The relevance of imaginary numbers
<http://www.math.toronto.edu/mathnet/answers/relevance.html>
<http://forum.swarthmore.edu/dr.math/problems/zakrzewski10.14.97.html>

Complex numbers in real life
<http://www.math.toronto.edu/mathnet/questionCorner/complexinlife.html>

Complex impedance in circuits
<http://hyperphysics.phy-astr.gsu.edu/hbase/electric/impcom.html>

Generation of fractals from complex numbers
<http://www.geocities.com/fabioc/>

▲▼▲

CONIC SECTIONS

In the third century B.C., the Greek mathematician Appollonius wrote a set of books dealing with what he called *conic sections*. He provided a visualization of *ellipses*, *hyperbolas*, and *parabolas* as intersections of planes with cones. Unlike ice cream cones, Appollonius's cone looked like two cones sharing a common vertex. The picture on the left shows the parabola that is formed by cutting the cones with a plane parallel to the slant of the cones. It took almost 2,000 years before applications of conic sections emerged in science and engineering, but they are now all around us. The middle picture shows a microwave antenna. The microwaves emerge from the transmitter outside of the reflector at its focus. The reflector concentrates the wave, as shown in the right-hand picture. Without the parabolic reflector, the waves would dissipate following the inverse square law. (See ***Inverse Square Function*** for more information.)

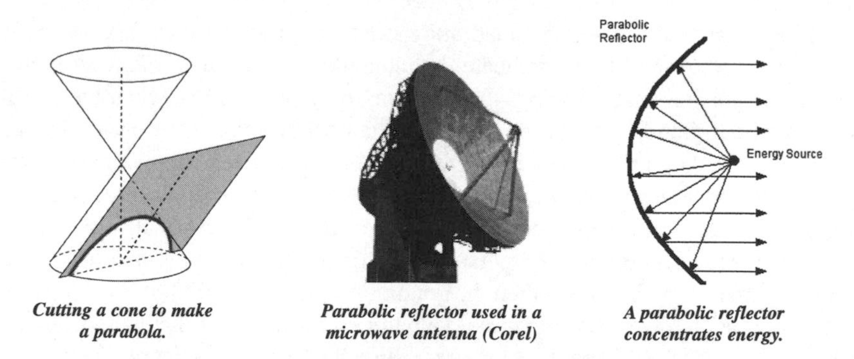

Cutting a cone to make a parabola.

Parabolic reflector used in a microwave antenna (Corel)

A parabolic reflector concentrates energy.

Various representations of the conic section called a "parabola."

The picture on the right shows how the waves from the energy source emerge from many directions. The energy source is positioned at the focus of the parabola. Once the rays hit the parabolic reflector, they are transmitted out in parallel direction. This concentrates the energy in one direction. For this reason, the parabolic shape is ideal for car headlights. It would also be ideal for television tubes were consumers not so demanding that picture screens be rectangular. Television tube manufacturers have to do some clever engineering to maximize the benefits of parabolic reflectors and still provide rectangular screens.

If the arrows are reversed in the right-hand drawing, then the parabolic reflector accumulates and concentrates energy from outside sources. For example, the dot for the energy source might represent a pipe containing water. Then the parabolic reflector can concentrate the sun's rays to heat the water as part of a solar heating system. Pipes in highly polished parabolic troughs can focus enough sunlight to heat an enclosed fluid as high as 750°F or turn water to steam. Hand-held parabolic reflectors that were invented for spying are available for sport and hobby activities such as bird watching. The parabolic reflector picks up weak sounds, such as distant bird calls, and focuses them on a microphone at the focal point.

Sometimes a diffuse view is important. Since they can provide almost 360° views, hyperbolic mirrors are used for security surveillance in buildings. The reflection in hyperbolic mirrors is from the convex side, rather than the concave side used for parabolic mirrors. This is what makes exterior mirrors on the passenger sides of cars show wider views and justify the warning, "Objects may be closer than they appear."

Parabolas appear in science and engineering. A hard-hit baseball flies off the bat in a parabolic path. The large cables strung between towers of a suspension bridge, such as the Golden Gate Bridge in San Francisco, form a parabola. Connection to the roadway of the bridge is important in shaping the large cables to parabolic shape. A telephone wire that curves because of its own weight is not a parabola, but is a *catenary*. If a heavy liquid like mercury is placed in a large can, and the can is spun, the surface of the liquid will form a paraboloid (every vertical cross section through the center of the can is a parabola). Parabolas are used in design and medical applications to determine smooth curves from three specified points in a solid or the image of a solid, such as the points provided in a medical CAT scan.

Ellipses are a oval conic section that look like squashed circles. They have two *foci* that act as centers of the ellipse. Hitting a ball from one focus on an elliptical pool table will result in a carom from the side of the table that sends the ball to the other focus. Rooms that have elliptical ceilings or shapes will reflect the sound of a pin dropping at one focal point so that it is audible many yards away at the other focal point. The Mormon Tabernacle in Salt Lake City and Statuary Hall in the U.S. Capitol in Washington, D.C., are two rooms that have remarkable acoustics because of their elliptical shapes.

Ellipses are an outcome of some common architectural techniques. The Romans invented the Groin Vault, the joining of two identical barrel (cylindrical) vaults over a square plan. The intersection of the vaults form ellipses that go diagonally to the corners of the square. Although the groin vault is common in ancient and medieval buildings, it is also found in modern structures such as the terminal building at the St. Louis Airport.

Johannes Kepler (1571–1630) revolutionized astronomy when he recognized that the motion of planets about the sun was elliptical and not circular. Working with the detailed planetary observations of Tycho Brahe (1546–1601), Kepler found some very slight errors in Brahe's figures for the circular orbit of Mars. He attempted to correct the values, but finally concluded the data was correct and that the orbit of Mars was elliptical with the sun at one of the focal points of the orbit. His verification of this for the other known planets of his time is known as "Kepler's first law." (See ***Variation***.)

Some comets, like Halley's comet, follow an elliptical path around the sun just like planets. Hence Halley's comet "returns" to earth's view on a regular basis. However, some comets appear to follow parabolic or hyperbolic paths. Once past the sun, they leave our solar system. These comets may have traced elliptical orbits at one time, but were thrown off trajectory by a gravitational encounter with a major planet such as Jupiter.

Many machines contain elliptical gears. These develop a nonuniform motion from a uniform power source. The momentary speedup or slowdown they produce is important in rotary shears, conveyers, motorcycle engines, and packaging machines.

Statisticians conceptualize plots of many variables on large numbers of subjects as elliptical swarms of points. By finding the axes of such swarms, they synthesize the information from many variables into important structural variables.

online sources for further exploration

Artistic views of conics

<http://www.xahlee.org/SpecialPlaneCurves_dir/ConicSections_dir/conicSections.html>

Conics in general

<http://www.iln.net/html_p/c/72782/62079/53803/53887.asp>

<http://www.kent.wednet.edu/KSD/KR/MATH/conic_sections2.html>

<http://nths.newtrier.k12.il.us/academics/math/Connections/curves/conics.htm>

<http://chs.osd.wednet.edu/nadelson/chsscimath/Conicsection2001/conic_section_creations.htm>

<http://www.ece.utexas.edu/projects/k12-fall98/14545/Group2/real.html>

Pictures of Appollonius's analysis

<http://www.sisweb.com/math/algebra/conics.htm>

<http://www.nsm.iup.edu/ma/gsstoudt/conics/conicsmma.html>

Explore conic sections dynamically

<http://www.keypress.com/sketchpad/java_gsp/conics.html>

<http://www.exploremath.com/activities/activity_list.cfm?categoryID=1>

Hyperbolic mirrors

<http://www.neovision.cz/prods/panoramic/h3b.html>

A video view of Statuary Hall in the U.S. Capitol

<http://www.discovery.com/news/picture/jul99/panoramas/javapano3.html>

Parabolic reflectors and antennas

<http://www2.gvsu.edu/~w8gvu/geo/geo.html>

How to build a parabolic reflector
<http://nths.newtrier.k12.il.us/academics/math/Connections/reflection/pararefl.htm>

Elliptical orbits
<http://csep10.phys.utk.edu/astr161/lect/history/kepler.html>
<http://www.bridgewater.edu/departments/physics/ISAW/PlanetOrbMain.html>

Elliptical gears
<http://www.cunningham-ind.com/ellipt.htm>
<http://www.hpceurope.com/vgb/archives/Avril00/Elliptiques.html>

▲▼▲

COUNTING

Businesses and government agencies often have a need to efficiently count the number of arrangements or possibilities with various combinations of numbers or options. For example, a car dealer may be interested in the number of car varieties that can be offered in order to persuade customers. If there are 9 different models, 6 different colors, and 2 types of interiors, there could be a total of $9 \times 6 \times 2$, or 108, different cars available. In this dealer's television advertisement you might hear, "Hurry, this weekend only. Come to our car dealership and view over 100 different styles of cars for sale. Don't miss out on this great opportunity!" The procedure of multiplying the number of possibilities for each option is called the *multiplication counting principle*.

State vehicle departments can determine the number of nonvanity license plates they have available by finding the product of the number of possibilities for each position on the plate. For example, if a state has three letters followed by three numbers, then the number of possible plates is $9 \times 10 \times 10 \times 26 \times 26 \times 26 = 9 \times 10^2 \times 26^3 = 15{,}818{,}400$. The first position will have 9 possible values, since it will represent any digit from 1 through 9. The second and third positions can hold 10 possible digits from 0 to 9. The fourth through sixth positions have 26 possibilities each, because they can contain any letter in the alphabet. If for some reason the state runs out of license-plate sequences, they can place numbers before letters to double the number of possibilities, since the order of letters and numbers is important on a license plate. Zip codes can be counted in a similar manner. There are five numbers in a zip code, so there is a total of $10 \times 10 \times 10 \times 10 \times 10 = 10^5 = 100{,}000$ possible zip codes. The United States Postal Service only uses 95,000 possibilities: 00001 to 95000.

Telephone numbers are counted in a similar way, but have more restrictions to the values in different positions. The first three digits are the area code. The area code must start with a digit from 2 to 9, because pressing 0 is a call to the operator, and pressing 1 is not allowed because it is associated with dialing a number outside an area code. Also, the area code cannot be 911, since that is an emergency number. Therefore there are a total of $8 \times 10 \times 10 - 1 = 799$ possible area codes. The local phone number has seven digits, with a three-digit prefix and four-digit suffix. In the prefix, the first digit cannot be 0 or 1 for the same reason mentioned earlier. Also, the prefix cannot use 555, because that is a dummy set of numbers used in entertainment media, such as movies and songs, except for the national information number, 555-1212. Therefore the prefix can have $8 \times 10 \times 10 - 1 = 799$ possible values. The suffix can have any four-digit number, which means there are $10 \times 10 \times 10 \times 10 =$ 10,000 possible values. Therefore, using the multiplication-counting principle, there is a total of $799 \times 799 \times 10,000 =$ 6,384,010,000 possible telephone numbers. That is an average of almost 25 numbers per person!

A lock manufacturer can determine the number of possible combinations to open its locks. If a dial lock has 60 numbers and requires three turns, then a total of $60 \times 60 \times 60 =$ 216,000 locks can be made. However, some lock companies do not want to have the same number listed twice, because dialing in different directions might end up being confusing. Therefore it might be more appropriate to create $60 \times 59 \times 58 =$ 205,320 lock combinations. The 59 in the second position means that there are 59 possible numbers available, because one number has been selected in the first position; and the 58 in the third position indicates that there are 58 possible numbers remaining, because one number has been selected in the first position and a different number has been selected in the second position. The product of three consecutive descending numbers is called a *permutation*. In this case, we would say that there are 60 permutations taken 3 at time, meaning that the counting accounts for the selection of three numbers out of a group of 60 in which the order of selection is important. Instead of writing the permutation as a product of a series of integers $n(n-1)(n-2) \bullet \ldots \bullet \ (n-r+1)$, it can be symbolized as $n\mathrm{P}r$, where n is the number of possibilities for the first selection, and r is the number of selections.

Some counting principles are based on situations in which the order of selection is not important, such as in selecting winning lottery balls. If 6 numbers are selected from a group of 40 numbers, it does not matter which number is pulled out of the machine first or last. After all the numbers are randomly drawn, the results are posted in numeric order, which is probably not the same order by which they were selected. For example, if the numbers are drawn in the order 35–20–3–36–22–28, and your ticket reads 3–20–22–28–35–36, then you are still the winner. When order of selection is not important, this type of counting principle is called a *combination* and can be symbolized as $n\mathrm{C}r$. The relationship between a *combination* and *permutation* is determined by the equation $n\mathrm{C}r = \frac{n\mathrm{Pr}}{n!}$ because there are $n!$ ways to arrange a group of n objects, where $n! = n(n-1)(n-2) \bullet \ldots \bullet 1$. In this case, there are $6! = 6 \times 5 \times 4 \times 3 \times 2 \times 1 = 720$ ways to rearrange 6 lottery balls with different numbers. Since the order of numbers is not important when reading the winning lottery number, there are ${}_{40}C_6$ possible numbers, or $\frac{40 \times 39 \times 38 \times 37 \times 36 \times 35}{6 \times 5 \times 4 \times 3 \times 2 \times 1} = 3,838,380$ combinations, to select in the lottery. In this type of lottery, the chance of winning would be 1 in 3,838,380.

online sources for further exploration

Adventure games, permutations, and spreadsheets
<http://archives.math.utk.edu/combinatorics/Combinatorics/AdvGame.html>

Counting techniques
<http://www.colorado.edu/education/DMP/activities/counting/>

English change-ringing of bells
<http://www.bris.ac.uk/Depts/Union/UBSCR/crinfo.html>

▲▼▲

DERIVATIVE. SEE RATES

▲▼▲

EQUATIONS

An *equation* is a relationship that places equal representation to different quantities, and is symbolized with an equals sign "=." All proportions are equations that are based on equal ratios. For example, Kepler's law states that the ratio of the cubed planetary distances from the sun is equal to the ratio of their squared period of revolutions around the sun, written as $\frac{d_1^3}{d_2^3} = \frac{p_1^2}{p_2^2}$. (See ***Proportions*** for more information about Kepler's law on planetary motion.) Functions can also be written as equations, because they relate two or more variables with an equals sign. For example, the vertical height, *h,* of an object is determined by the quadratic equation $h = -0.5gt^2 + v_0t + h_0$, where *g* is the acceleration due to earth's gravity (9.8 m/sec^2), v_0 is the initial vertical velocity, and h_0 is the initial height of the object (see ***Quadratic Functions*** for more information about the derivation and use of this equation). There are many other types of relationships besides proportions and functions that can be represented in the form of an equation. Some equations are bizarre and have multiple variables, making them interesting to study or purposeful to use. Other equations simply do not belong to a family of functions that is commonly studied in high school mathematics.

In 1622, chemist Robert Boyle showed that the product of the pressure, *p*, and volume, *v*, of the same amount of gas at a fixed temperature was constant. The equation to represent this relationship is $p_1v_1 = p_2v_2$, where the subscript notation represents the gas at different times. This formula indicates that as the pressure increases, the volume of the gas will decrease, and vice versa. For example, when diving under water, the amount of pressure in your ear sockets will increase, causing the amount of space to decrease until your ears "pop." The amount of space in your lungs also decreases when you are underwater, making it more difficult to breath when scuba diving. One way to visualize this effect is to bring a closed plastic container of soda onto an airplane, and then notice the change in its shape during takeoff and descent due to varying pressures in the earth's atmosphere at different altitudes. If temperature, t, and quantity of gas in moles, n, vary, then the equation can be extended to the ideal gas law, which is $pv = nrt$, where r is the universal gas constant equal to 0.082 (atm L)/(mol K).

The escape velocity of an object represents the speed at which it must travel in order to escape the planet's atmosphere. On earth, it is the speed at which a rocket or shuttle needs in order to break the gravitational pull of the planet. The equation that relates the escape velocity, v_e, to the mass, M, and radius, R, of a planet is approximately $v_e^2 = (1.334 \times 10^{-10})(M/R)$. The equation is based on finding the moment when the kinetic energy, $0.5mv_e^2$, of the rocket exceeds its potential energy that is influenced by the earth's gravitational pull, GMm/R, where G is a gravitational constant, 6.67×10^{-11}, and m is the mass of the rocket. Setting these two relationships equal to one another, $0.5mv_e^2 = GMm/R$, sets up a situation that determines the velocity at which the kinetic and potential energy of the rocket are the same. An m on both sides of the equation cancels and the equation simplifies to $v_e^2 = (1.334 \times 10^{-10})(M/R)$. The mass of the earth is $5.98 \star 10^{24}$ kg, and has a radius of 6,378,000 m. This means that a rocket needs to exceed 11,184 meters per second to fly into space. That is almost 25,000 miles per hour!

Equations involving the sum of reciprocals exist in several applications. For instance, the combined time to complete a job with two people, T_c, can be determined by the equation $1/T_1 + 1/T_2 = 1/T_c$, where T_1 and T_2 represent the time it takes two different individuals to complete the job. This equation is based on the equation $P = RT$, where P is the worker's productivity, R is the worker's rate, and T is the worker's time on the job. Since two workers complete the same job, they will have the same productivity level. This means that the two workers' productivity can be represented by the equations $P = R_1T_1$ and $P = R_2T_2$. The productivity for both workers is based on a combined rate and different time, represented with $P = (R_1 + R_2)T_c$. Substituting $R_1 = \frac{P}{T_1}$ and $R_2 = \frac{P}{T_2}$ makes the equation $P = (\frac{P}{T_1} + \frac{P}{T_2})T_c$. Dividing both sides by T_c and canceling the productivity variable leaves the end result, $\frac{1}{T_1} + \frac{1}{T_2} = \frac{1}{T_c}$.

Suppose an experienced landscaper can trim bushes at a certain house in 3 hours, and a novice takes 5 hours to complete the same job. Together, they will take 1 hour, 52 minutes, and 30 seconds to complete the task, assuming that they are working at the same productivity level (i.e., they are not distracting each other's performance by chatting). This result was determined by solving the equation $\frac{1}{3} + \frac{1}{5} = \frac{1}{T_c}$. If both sides of the equation are multiplied by the product of the fraction's denominators, or $15T_c$, the equation can be rewritten as $5T_c + 3T_c = 15$. After combining like terms and dividing both sides of the equation by 8, the solution will be $T_c = \frac{15}{8}$, which translates to 1 hour, 52 minutes, and 30 seconds. Reciprocal equations exist in other applications as well. The image formed by a converging or diverging lens can be located with the equation $\frac{1}{D_i} + \frac{1}{D_o} = \frac{1}{F}$, where D_i is the distance from the lens to the image, D_o is the distance from the object to the lens, and F is the focal distance of the lens.

Sports statistics involve unusual equations. The NFL quarterback rating is a computation that measures the effectiveness of a player based on his number of touchdowns (t), interceptions (i), attempts (a), completions (c), and passing yards (p). The equation that determines the quarterback rating, r, is

$$r = (500c + 25p + 2000t + 12.5a - 2500i)/(6a).$$

Notice that interceptions are weighted so that the rating decreases by more than the value of a touchdown, and that touchdown passes are weighted four times as much as a completion. This equation is proportioned so that the average quarterback will have a rating near 100, according to historical performances in the league. This equation is based on NFL statistics and needs to be adjusted for other football leagues, since the fields and rules are slightly different. For example, scoring in the Arena Football League occurs more often, since the field is only 50 yards long, compared to 100 yards in the NFL.

online sources for further exploration

Arena Football League quarterback rating
<http://www.tampastorm.com/features/QBrate/>

Robert Boyle and his data
<http://dbhs.wvusd.k12.ca.us/GasLaw/Gas-Boyle-Data.html>

Boyle's Law and absolute zero and Cartesian diver and Model of Lung
<http://chemlearn.chem.indiana.edu/demos/Boyle.htm>

Burning rate of stars
<http://www.phys.uri.edu/~chuck/ast108/notes/node76.html>

Calculate the escape velocity
<http://www-star.stanford.edu/projects/mod/ad-escvel.html>

Euler's formula and topology
<http://www.nrich.maths.org/mathsf/journalf/dec00/art1/index.html>

Ideal gases
<http://library.thinkquest.org/3616/chem/gas.htm>

Image forming by a lens
<http://www.lightlink.com/sergey/java/java/clens/index.html>
<http://www.lightlink.com/sergey/java/java/dlens/index.html>

Orbit simulation
<http://observe.ivv.nasa.gov/nasa/education/reference/orbits/orbit3.html>

Quarterback rating system
<http://user.cybrzn.com/~koz/rating.htm>
<http://www.primecomputing.com/javaqbr1.htm>

Seventeen proofs of Euler's formula
<http://www.ics.uci.edu/~eppstein/junkyard/euler/>

Universal law of gravitation
<http://csep10.phys.utk.edu/astr161/lect/history/newtongrav.html>

▲▼▲

EXPECTED VALUE

The *expected value* of a variable is the long-run average value of the variable. Expected value can also be viewed as the average value of a statistic over an infinite number of samples from the same population.

Studies of expected value emerged from problems in gambling. How much is a lottery ticket worth? Consider a lottery run by a service organization: a thousand tickets are offered at a dollar each; first prize is $500; there are two second prizes of $100; and the remaining income from ticket sales is designated for charity. There are three probabilities in this problem: The probability of having the first-prize ticket is 1 out of 1,000, or 0.001; the probability of a second place ticket is 0.002; and the probability of winning nothing is 0.997. The average of prizes weighted with corresponding probabilities gives the expected winning for a ticket: $500 \bullet 0.001 + 100 \bullet 0.002 + 0 \bullet 0.997 = 0.700$. The expected-prize value for one of these lottery tickets is $0.70. Since the ticket costs a dollar, the expected loss on a ticket is $0.30. For this model to hold, one must assume that a ticket would be purchased from many such lotteries. This assumption is met by state lotteries that sell millions of tickets or Las Vegas slot machines, which are played millions of times each day. Neither lotteries nor slot machines are *fair* games. The expected net winning for each ticket in the lottery or each play of a slot machine is a negative number. This indicates that these games of chance represent a long-term loss for the regular gambler.

The concept of weighting costs by probabilities is used in finance, investing, insurance, industrial decision-making, and law to determine expected values. Bankers and investors use several indicators based on expected value. One example is *expected return*, an expected value on a risky asset based on the probability distribution of possible rates of return that might include U.S. Treasury notes, stock-market indices, and a risk premium. Industrial decision-making uses expected values to compute projected costs of different options. For example, an oil company may hold property that it may choose for oil drilling, hold for later drilling, or sell. Each of these options is associated with costs. The company can compute probabilities based on past experience for each cost. They then compare the expected values and choose the option that has the least expected cost. A controversial industrial use of expected value occurred in the 1970s with the design of the Ford Pinto automobile. It had a gas tank that was likely to explode when the car suffered a rear-end collision. The Ford Motor Company computed expected costs of improving the Pinto gas tank versus the expected costs of settling lawsuits resulting from deaths in Pinto explosions. The latter value was the lesser, so Ford executives chose to omit gas-tank improvements.

Law firms can use expected values to determine whether or not a client should continue a suit, settle without a trial, or go to trial. Experience with similar lawsuits provides the probabilities. The cost of litigation and the potential awards provide the estimates of net "winnings." If the expected value of the net winnings in a trial is negative, the law firm should advise the client to drop the suit or accept a settlement.

Ecologists have used expected value to estimate water supplies in the Great Plains based on probability and volume estimates of soil moisture, rain, and consumption by humans, industry, agriculture, and natural vegetation. The military uses expected values in conducting "war games." Costs in military operations, loss of life, and destruction of property are associated with probabilities to compare the expected values of different strategies.

The French mathematician and philosopher Blaise Pascal (1623–1662) provided one of the earliest and most intriguing uses of expected value. In what is now called "Pascal's wager," he argued that probabilities and payoffs associated with belief in God versus not believing in God would result in an expected value that supported belief in God. Almost 400 years later, Pascal's assumptions and arguments are still debated by theologians and philosophers.

online sources for further exploration

The cereal box problem
<http://www.mste.uiuc.edu/reese/cereal/intro.html>

Contest odds
<http://silver.sdsmt.edu/~rwjohnso/module7.htm>

Determination of the decision-maker's utility function
<http://ubmail.ubalt.edu/~harsham/opre640a/partix.htm#rutility>

The "dummies guide" to lottery design
<http://www.parliament.the-stationery-office.co.uk/pa/cm199900/cmselect/cmcumeds/958/01111622.htm>

Life-expectancy data
<http://www.iihe.org/information/Databook1996/T49_LifeExpectancyAgeRaceSex.htm>

▲▼▲

EXPONENTIAL DECAY

Exponential decay can be observed in the depreciation of car values, the half-life of elements, the decrease of medication in the blood stream, and the cooling of a hot cup of coffee. The general exponential equations that define exponential growth, such as the financial model for principal after compound interest is applied, $A = P(1 + \frac{r}{n})^{nt}$, and the general models for exponential growth such as $y = ab^x$ can be used to describe losses over time for values of b that are between 0 and 1. The changes that are made to the models may involve changing the base from a number greater than one (growth) to a number less than one (decay), or leaving the base alone and allowing the power to be negative.

The term "decay" comes from the use of exponential functions to describe the decrease of radioactivity in substances over time. The law of radioactive decay states that each radioactive nuclear substance has a specific time known as the *half-life*, during which radioactive activity diminishes by half. Some radioactive substances have half-lives measured in thousands to billions of years (the half-life of uranium-238 is 4.5 billion years), and some in fractions of a second (muons have a half-life of 0.00000152 seconds). The way in which radioactivity is measured varies from substance to substance. Uranium-238 decays into lead, so the proportions of lead and uranium-238 in a sample can be used to determine the amount of decay over time. The law of decay is stated as $A_R = A_o(\frac{1}{2})^{t/h}$, where A_o is the amount of radioactive substance at the start of the timing, h is the half-life time period, and A_R is the amount remaining after t units of time. In this format, the base of the exponential equation is $\frac{1}{2}$, clearly a number less than one. It can also be stated with a base larger than one if the exponent is negative, as in $A_R = A_o(2)^{-t/h}$. The basic shape of the graph of exponential decay is shown in the plot below. One hundred grams of substance with half-life of 24,000 years is followed for 100,000 years. At the end of 24,000 years, 50 grams of the radioactive substance are left in the sample. At the end of 48,000 years, 25 grams are left, and at the end of 72,000 years, 12.5 grams. The formula that describes this model is $A = 100(\frac{1}{2})^{t/24{,}000}$.

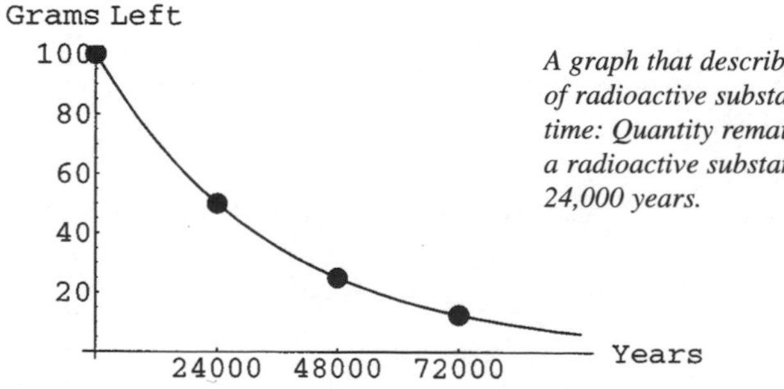

A graph that describes an exponential decay of radioactive substance as a function of time: Quantity remaining of 100 grams of a radioactive substance with half-life of 24,000 years.

Exponential decay models are also written using base e. The equation $A = 100e^{-kt}$, where $k = \frac{ln2}{24{,}000}$ is the same equation plotted in the graph.

Radiocarbon dating of animal or plant remains that are thousands of years old is based on the radioactive isotope carbon-14, which has a half-life of 5,700 years. Carbon-14 is constantly produced in the earth's atmosphere through the absorption of radiation from the sun. When living organisms breathe or eat, they ingest some carbon-14 along with ordinary carbon. After an organism dies, no more carbon-14 is ingested, so the age of its remains can be calculated by determining how much carbon-14 is left.

Exponential decay in prices is called *depreciation.* Some types of depreciation used in accounting are linear. For example, tax law permits a business to depreciate 20 percent of the original cost of computer equipment for each of five years. Market prices, however, do not follow a linear pattern. Automobiles typically depreciate rapidly during the first year, and then less rapidly during each subsequent year. The used-car prices for one popular automobile that sold for \$27,000 when new are given by $P = 27,000(0.83)^t$, where t is the number of years after purchase. In this case, the automobile lost 17 percent of its value each year.

Inflation problems can be viewed as growth problems (increases in prices) or as drops in the value of currency. For example, the purchasing power of the dollar dropped by 7.2 percent per year during the 1970s. The purchasing power of \$100 is given by $P = 100(1 - 0.072)^t = 100(0.928)^t$, where t is the number of years after 1970.

Concentrations of a medication that are carried in the bloodstream often follow an exponential decay model. Such drugs are said to have half-lives. Each day you replace about 25 percent of the fluids in your blood. If you are taking a medication that depends on the bloodstream for circulation, then 25 percent of the dose is lost as you replace fluids. A person who takes one pill containing 20 mg of medicine will have about 15 mg (75 percent of 20 mg) in his or her body one day later, and 11.25 mg (75 percent of 15 mg) two days later, and so on. The half-life for this drug can be found by solving the equation $\frac{1}{2} = 0.75^t$, or $t \approx 2.4$ days.

Some drugs do not follow an exponential decay pattern. Because alcohol is metabolized by humans, the quantity of alcohol in the bloodstream after ingestion will show a linear decrease rather than exponential decay.

For the many drugs and steroids that have half-lives, the drop off in drug concentration decreases less rapidly over time. Therefore it is possible to measure the quantity of the drug in the body long after ingestion. This means that users of illegal or dangerous drugs will have traces of the drugs remaining in their bloodstreams for many days. Sensitive drug tests, such as those used on Olympic athletes, can pick up indications of banned drugs used within two weeks or more of the testing, depending on the half-life of the substance.

If you pour a cup of hot coffee, the temperature will drop off quickly, then the coffee will remain lukewarm for a long while. *Newton's law of cooling* states that the rate at which the temperature drops is proportional to the difference between the coffee temperature and the room temperature. The formula for the coffee temperature after t minutes is $T = T_r + (T_o - T_r)e^{-kt}$, where T_o is the initial temperature of the coffee, T_r is the room temperature, and k is a constant depending on the type of cup. Hence cooling is an exponential decay situation. (See ***Asymptote***.)

Medical examiners use a version of Newton's law of cooling to determine the time of death based on the temperature of a corpse and ambient temperature at the murder scene.

online sources for further exploration

Journal of Online Mathematics and Its Applications
<http://www.joma.org/vol1-2/modules/macmatc5/exponential_decay_module.html>

Carbon dating
<http://www.c14dating.com/>
<http://www.cs.colorado.edu/~lindsay/creation/carbon.html>

Cooling
<http://mvhs1.mbhs.edu/mvhsproj/cooling.html>
<http://members.tripod.com/fix_it_quick/mathisu.html>
<http://www.aw.com/ide/Media/JavaTools/nlhcrate.html>

Nuclear medicine
<http://www.math.bcit.ca/examples/ary_11_4/ary_11_4.htm>

Radioactive decay
<http://www.joma.org/vol1-2/modules/macmatc5/exponential_decay_module.html>
<http://pass.maths.org.uk/issue14/features/garbett/index.html>

The RC circuit
<http://www.math.bcit.ca/examples/ary_7_4/ary_7_4.htm>

Used car prices
<http://www.edmunds.com/used/>
<http://www.kbb.com/kb/ki.dll/kw.kc.bz?kbb&&688&zip_ucr;1409&>

▲ ▼ ▲

EXPONENTIAL GROWTH

Exponential growth situations are based on repeated multiplication. A classic example was the growth of the rabbit population in Australia. English wild rabbits are not native to Australia, but were introduced by Thomas Austin of Winchelsea, Victoria, onto his property in 1859. Australia provided an ideal environment for the rabbits, with plenty of food and no predators, so the population grew rapidly. By 1910, rabbits had become a plague, driving out many of the native species across Australia. They destroyed farming areas, caused severe erosion, and ruined grazing areas for sheep. During the thirty years after their introduction, the rabbit population doubled every six months. The table and graph below show the approximate number of rabbits in Australia for each six-month period after the introduction of Mr. Austin's original 24 rabbits.

The growth of the rabbit population in Australia from 1859–1864 modeled by an exponential growth equation.

year	*six-month periods (x)*	*rabbit population (y)*
1859	0	24
	1	$24 \bullet 2 = 48$
1860	2	$24 \bullet 2 \bullet 2 = 96$
	3	$24 \bullet 2 \bullet 2 \bullet 2 = 192$
1861	4	$24 \bullet 2^4 = 384$
	5	$24 \bullet 2^5 = 768$
1862	6	$24 \bullet 2^6 = 1{,}536$
	7	$24 \bullet 2^7 = 3{,}072$
1863	8	$24 \bullet 2^8 = 6{,}144$
	9	$24 \bullet 2^9 = 12{,}288$
1864	10	$24 \bullet 2^{10} = 24{,}576$

The equation for the number of rabbits is $y = 24 \bullet 2^x$. Because the independent variable x is in an exponent, the equation describing the rabbit population growth is called an *exponential model.* The base 2, which represents doubling, is the growth factor. Growth factors greater than one create curves similar to the rabbit-population curve. When the growth factor is less than one, the curve will decrease (see ***Exponential Decay***).

Exponential growth models are used extensively in the world of finance. Investments of money in a certificate of deposit (CD), for example, require the customer to invest a certain amount of money (principal) for a specified time. The bank issuing the CD will specify an annual yield, a yearly interest rate that will be added to the principal each year. The interest becomes part of the princi-

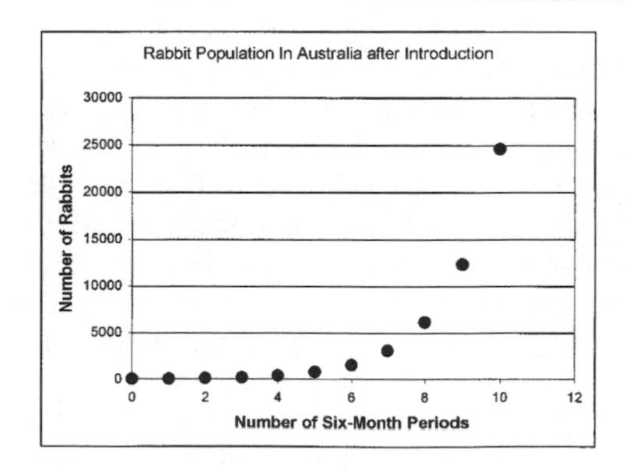

A graph depicting growth of the rabbit population in Australia from 1859–1864.

pal held for the customer. The return addition of interest payments to the principal so that the interest amount can earn interest in later years is called *compound interest*. The growth factor in compound-interest problems is 1 plus the annual yield. So an investor who buys a \$5,000 CD advertised at 6.5 percent annual yield will receive $5000(1 + .065)^x$ after x years. After three years, this CD would be valued at $5000(1.065)^3$ = \$6,039.75. Banks may choose to compound interest more frequently. The banking version of the exponential growth formula is $A = P(1 + r/n)^{nt}$, where A is the amount at the end of t years, P is the starting principal, r is the stated interest rate, and n is the number of periods per year that interest will be compounded. A typical CD will have interest compounded each quarter. Financial institutions can offer more-frequent compounding, such as monthly or daily. Some even offer continuous compounding, which has the formula $A = Pe^{rt}$, where A is the value of the investment at time t, P is the initial principal, r is the interest rate, and $e \approx 2.7183$. For a given interest rate, more frequent compounding yields a higher return, but that return does not increase dramatically as the compounding period moves from months to days to continuous. Because the number of compounding periods can affect the rate of return on an investment, federal law requires financial institutions to state the annual yield as well as an interest rate so that consumers can make easier comparisons among investment opportunities.

Benjamin Franklin was one of the pioneers in the use of exponential growth models for money and population. In 1790, Franklin established a trust of \$8,000. He specified that his investment should be compounded annually for 200 years, at which time the funds should be split evenly between the cities of Philadelphia and Boston, and used for loans to "young apprentices like himself." Franklin anticipated that the fund would be worth \$20.3 million after 200 years if the annual yield averaged 4 percent. However, the annual yield averaged about 3.4 percent, so \$6.5 million was in the fund when it was dispersed to the two cities in 1990.

Franklin established the practice of studying the American population by using exponential growth. He recognized that the warning of the Englishman Thomas Malthus (1766–1834) that population under exponential growth would outstrip food sources might apply to the new country of the United States. Franklin urged that the growth of states and the entire country be tracked each year. Some historians contend that President Lincoln used exponential growth models 70 years after Franklin's recommendation. Lincoln used censuses from 1790 to 1860 to predict that the population of the United States would be over 250 million in 1930. The population did not reach this figure until 1990. This shows that exponential functions can describe situations only as long as the growth factor remains constant. There are many factors such as economics, war, and disease that can affect the rate of population growth.

When the Center for Disease Control identifies a new epidemic of flu, exponential growth functions describe the numbers of early cases of infection quite well. A good definition of epidemic is a situation in which cases of disease increase exponentially. However, as people build up immunization, the disease cannot continue exponential growth, and other models become more appropriate. (See ***Logistic Functions***.)

The federal government keeps close tab on exponential growth situations that can or may harm the U.S. economy. Inflation is the growth in prices over time. One measure of inflation is the Consumer Price Index (CPI), which provides averages of what standard goods and services would cost each year. In the United States, what cost $100 in 1980 would cost $228.69 in 2000. The value of a dollar was therefore less in 2000 than it was in 1980. This corresponds to a yearly increase in costs of about 4.2 percent. This can be checked with the exponential growth calculation $100(1 + 0.042)^{20} \approx \$227.70$. This inflation is not a serious national problem if wages and salaries increase at the same rate. It becomes a crisis if the costs of goods and services increase at too high a percentage. There was a time during the last twenty years in which the inflation rate in Brazil reached 80 percent *per month*! Using the exponential growth equation, that means that what cost $100 at the start of the year would cost $100(1 + 0.80)^{12} \approx \$1{,}157$ at the end of the year.

Exponential growth is an issue in studies of the environment. From 1950 through 1970, it appeared that world oil production was increasing exponentially at a rate of 7 percent per year to meet the growing worldwide demand. Could that continue? Because it is harder to find previously undiscovered oil deposits, oil production has not increased exponentially since 1970. Some scientists contend that the carbon dioxide content in the upper atmosphere is increasing exponentially. There are few dangerous effects in the early stages of the growth, but as the amount of atmospheric CO_2 leaps ahead, serious changes such as global warming will disrupt life on earth.

Exponential growth models are the basis of many scams, such as the chain letter. A chain letter offers the promise of easy money. One letter might have five names at the end of it. "Send $10 to the first name on the list. Remove that name and put your name on the bottom of the list. Send copies of the new letter to five people." If you and everyone else does this, the person at the top of the list would receive $6,250. However, by the time your name came up on top of the list, 1,953,125 people would have had to pass on the chain letter after it had been initiated. In three more stages, the letter would have to be continued by more people than there are in the United States. Because the number of contributors to the letter must grow exponentially, the only people who benefit from a chain letter are those who start them. The U.S. Postal Code prohibits chain letters. However, variants of chain letters that don't ask for money have been popular via email. Because these letters ask the recipient to send copies of the letter to all people in their computer address books, the number of these messages increases very rapidly and can clog disk storage and communication links.

There are several other ways in which exponential growth appears in financial deceits. An entrepreneur will advertise franchises for selling some product. For payment of a franchise fee, such as $1,000 or $5,000, the franchisee obtains the rights to sell the product in a certain area. Up to that point, everything is legal. But some frauds depend on the franchisees selling further franchises, with everyone already in the business sharing some of the franchise fees. In this type of scheme, millions of dollars can come to the originators, even if none of the product is ever sold. The people who pay franchises late in the scheme lose all their money. When all operations are based on money from new investors rather than goods or services, the fraud is called a "Ponzi scheme."

online sources for further exploration

Population changes

<http://www.ea.gov.au/biodiversity/invasive/pests/rabbit.html>
<http://www.learner.org/exhibits/dailymath/population.html>
<http://www.joma.org/vol1-2/modules/macmatc5/exponential_growth_module.html>
<http://www.math.montana.edu/frankw/ccp/modeling/discrete/snooping/learn.htm>

Savings, credit, and compound interest

<http://www.learner.org/exhibits/dailymath/savings.html>
<http://www.richmond.edu/~ed344/webunits/math/banking3.html>
<http://www.math.toronto.edu/mathnet/questionCorner/mortgage.html>

Inflation rates and calculators

<http://www.westegg.com/inflation/>
<http://woodrow.mpls.frb.fed.us/economy/calc/cpihome.html>
<http://www.hec.ohio-state.edu/cts/osue/cpidist.htm>

Chain letters and scams

<http://hoaxbusters.ciac.org/HBHoaxInfo.html#what>
<http://www.usps.gov/websites/depart/inspect/chainlet.htm>
<http://www.chainletters.org/>
<http://home.nycap.rr.com/useless/ponzi/>
<http://www.bosbbb.org/lit/0052.htm>

Food technology

<http://www.math.bcit.ca/examples/ary_2_4/ary_2_4.htm>

Internet growth data

<http://www.mit.edu/people/mkgray/net/internet-growth-summary.html>

Pricing diamond rings

<http://exploringdata.cqu.edu.au/dia_asn.htm>

The US national debt clock

<http://www.brillig.com/debt_clock/>

▲ ▼ ▲

FIBONACCI SEQUENCE

The infinite sequence 1, 1, 2, 3, 5, 8, 13, 21, 34, . . . is called the *Fibonacci sequence* after the Italian mathematician Leonardo of Pisa (ca.1175–ca.1240), who wrote under the name of Fibonacci. The sequence starts with a pair of ones, then each number is the sum of the two preceding numbers. The formula for the sequence is best written recursively (first formula below), rather than the explicit formula on the right.

$$\begin{cases} a_1=1 \\ a_2=1 \\ a_n=a_{n-1}+a_{n-2} \end{cases} \qquad t_n = \frac{\left(\frac{1+\sqrt{5}}{2}\right)^2 - \left(\frac{1-\sqrt{5}}{2}\right)^2}{\sqrt{5}}.$$

recursive ***explicit***

Fibonacci established a thought experiment about counts of animals over generations, and can be described in terms of the family line of honey bees. A male bee develops from an unfertilized egg—hence has only a mother. Female bees develop from fertilized eggs; therefore female bees have a father and mother. How many ancestors does a male bee have? The male bee has one mother. The mother has a mother and a father. So the male bee has one ancestor at the parent generation. He has two ancestors at the grandparent generation. If you work out the great-grandparent generation, you will find that there are three ancestors. A full picture of the family tree for the bee going back to great-great-great grandparents will show that the generation counts are 1, 2, 3, 5, 8, 13.

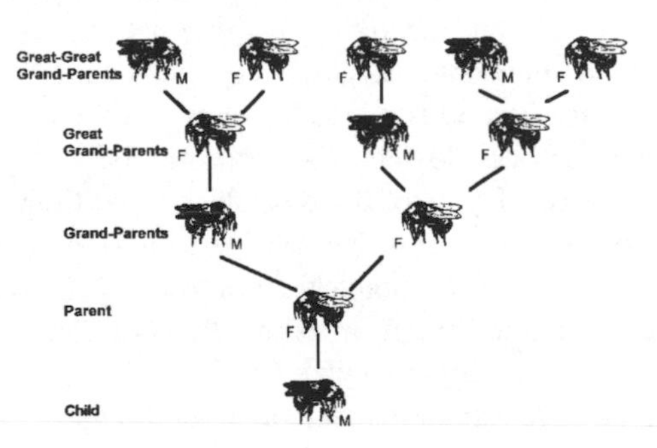

The family tree for a male bee.

Placing the male bee at the beginning of the sequence (starting generation) gives 1, 1, 2, 3, 5, 8, and so on. If you repeat the argument with a female bee, you will also get a Fibonacci sequence starting with 1, 2, 3, 5, 8, ... The sequence has been shown to have remarkable mathematical properties and some surprising connections to events outside of mathematics. Eight hundred years after Fibonacci's publication of the sequence, an organization and journal, the *Fibonacci Quarterly*, are devoted to exploring new discoveries about the sequence.

The ratios of consecutive terms of the Fibonacci sequence $\left(\frac{a_n}{a_{n-1}}\right)$ produce a sequence $1, 2, 1.5, 1.\bar{6}, 1.6, 1.625, \ldots$ which converges to the *golden ratio* $\frac{1+\sqrt{5}}{2} \approx 1.61803$. If a sequence of squares is built up from two initial unit squares (left-hand picture below), the vertices provide links for tracing a logarithmic spiral (middle picture). The spiral (which expands one golden ratio during each whole turn) appears in the chambered nautilus (right-hand picture).

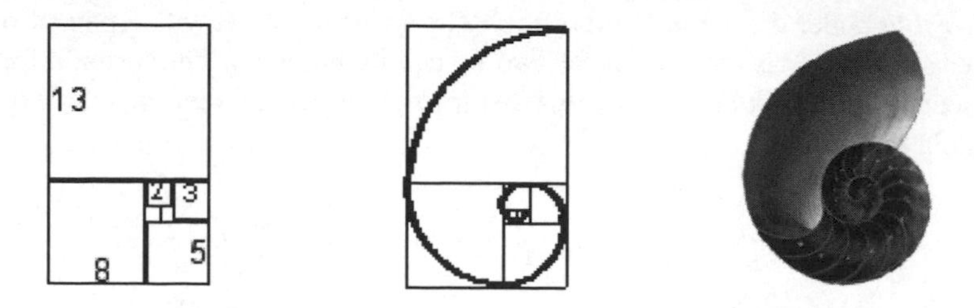

The Fibonacci numbers and their connection to the golden ratio in a chambered nautilus.

The Fibonacci numbers appear in the branching of plants, and counts of spirals in sunflower seeds, pine cones, and pineapples. In one particular variety of sunflower, the florets appear to have two systems of spirals, both beginning at the center. There are fifty-five spirals in the clockwise direction, and thirty-four in the counterclockwise one. The same count of florets in a daisy show twenty-one spirals in one direction and thirty-four in the other. A pine cone has two spirals of five and eight arms, and a pineapple has spirals of five, eight, and thirteen. The spiral also appears in animal horns, claws, and teeth.

On many plants, the number of petals on blossoms is a Fibonacci number. Buttercups and impatiens have five petals, iris have three, corn marigolds have thirteen, and some asters have twenty-one. Some species have petal counts that may vary from blossom to blossom, but the average of the petals will be a Fibonacci number. Flowers with other numbers of petals, such as six, can be shown to have two layers of three petals, so that their counts are simple multiples of a Fibonacci number. In the last few years, two French mathematicians, Stephane Douady and Yves Couder, proposed a mathematical explanation for the Fibonacci-patterned spirals in nature. Plants develop seeds, flowers, or branches from a meristem (a tiny tip of the growing point of plants). Cells are produced at a constant rate of turn of the meristem. As the meristem grows upward, the cells move outward and increase in size. The most efficient turn to produce seeds, flowers, or branches will result in a Fibonacci spiral.

In 1948, R. N. Elliott proposed investment strategies based on the Fibonacci sequence. These remain standard tools for many brokers, but whether they are a never-fail way of selecting stocks and bonds is open to debate. Some investors think that when Elliott's theories work, it is because many investors are using his rules, so their effects on the stock market shape a Fibonacci pattern. Nevertheless, a substantial number of brokers use Elliott's Fibonacci rules in determining how to invest.

In computer science, there is a data structure called a "Fibonacci heap" that is at the heart of many fast algorithms that manipulate graphs. Physicists have used Fibonacci sequences to study quantum transport through Fibonacci lattices and radiation paths through the solar system.

online sources for further exploration

An absolutely huge collection of information about the Fibonnaci sequence
<http://www.ee.surrey.ac.uk/Personal/R.Knott/Fibonacci/fib.html>

Computer art based on Fibonacci numbers
<http://www.moonstar.com/~nedmay/chromat/fibonaci.htm>

Fibonacci numbers and the golden section
<http://www.ee.surrey.ac.uk/Personal/R.Knott/Fibonacci/fib.html>

Fibonacci numbers and their application in trend analysis
<http://library.shu.edu/HafnerAW/awh-th-fibonacci-num.htm>

Fibonacci spirals
<http://www.moonstar.com/~nedmay/chromat/fibonaci.htm>

A psychic encounter with Fibonacci numbers
<http://www.telepath.com/novelty/nbart1.html>

Scott's phi page
<http://www.germantownacademy.org/academics/us/Math/Geometry/stwk98/SCOT-TRK/Index.htm>

Trader's corner
<http://www.optioninvestor.com/traderscorner/070501_1.asp>

▲▼▲

IMAGINARY NUMBERS. See COMPLEX NUMBERS

▲▼▲

INTEGRATION

Integration is used to determine a total amount based on a predictable rate pattern, such as a population based on its growth rate, or to represent an accumulation of something such as volume in a tank. It is usually introduced in calculus, but its use and computation can be performed by many calculators or computer programs without taking calculus. Understanding the utility of an integral does not require a background in calculus, but instead a conceptual understanding of rates and area.

Many realistic applications of integration that occur in science, engineering, business, and industry cannot be expressed with simple linear functions or geometric formulas. Integration is powerful in such circumstances, because there is not a reliance on constant rates or simple functions to find answers. For example, in many algebra courses, students learn that distance = rate $\times$ time. This is true only if the rate of an object always remains the same. In many real-world instances, the rate of an object changes, such as the velocity of an automobile on the road. Cars speed up and slow down according to traffic signals, incidents on the road, and attention to driving. If the velocity of the car can be modeled with a nonlinear function, then an integral could help you represent the distance as a function of time, or tell you how far the car has moved from its original position, even if the rate has changed.

A definite integral of a function $f(t)$ is an integral that finds a value based on a set of boundaries. A definite integral can help you determine the total production of textiles based on a specific period of time during the day. For example, suppose a clothes manufacturer recognized that its employees were gradually slowing down as they were sewing clothes, perhaps due to fatigue or boredom. After collecting data on a group of workers, the manufacturer determined that the rate of production of blue jeans, f, can be modeled by the function $f(t) = 6.37e^{-0.04t}$, where t is the number of consecutive hours worked. For the first two hours of work, an expected production amount can be determined by the definite integral, written as $\int_0^2 6.37e^{-0.04t}dt$.

On a graph in which $f(t)$ describes a rate, the definite integral can be determined by finding the area between f and the t axis.

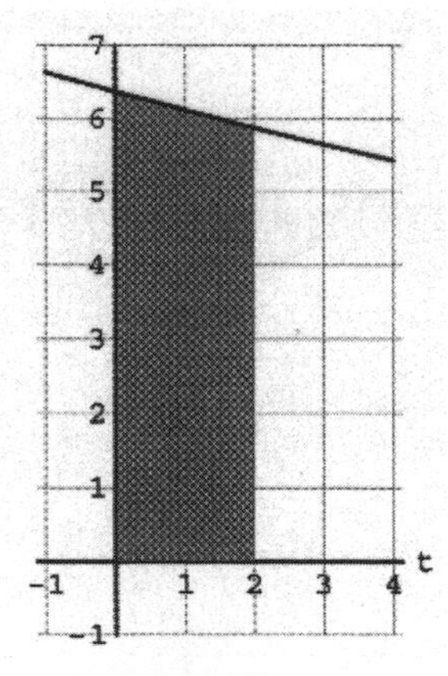

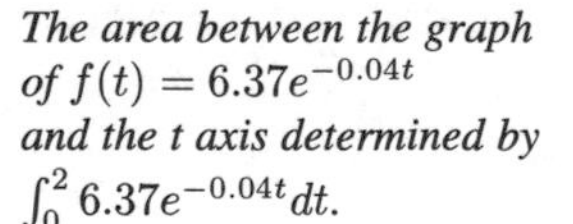
The area between the graph of $f(t) = 6.37e^{-0.04t}$ and the t axis determined by $\int_0^2 6.37e^{-0.04t}dt$.

In the case of producing blue jeans for the first two hours of work, the area between $f(t) = 6.37e^{-0.04t}$ and the t axis on the interval [0,2] is equal to approximately 12.24 pairs of jeans. In an eight-hour workday, the last two hours of work production from an employee would be represented by $\int_6^8 6.37e^{-0.04t}dt$, which equals approximately 9.63 pairs of blue jeans. Notice that the area on the graph is much lower in this interval (the dark solid region), than from 0 to 2 hours of work (the light shaded region).

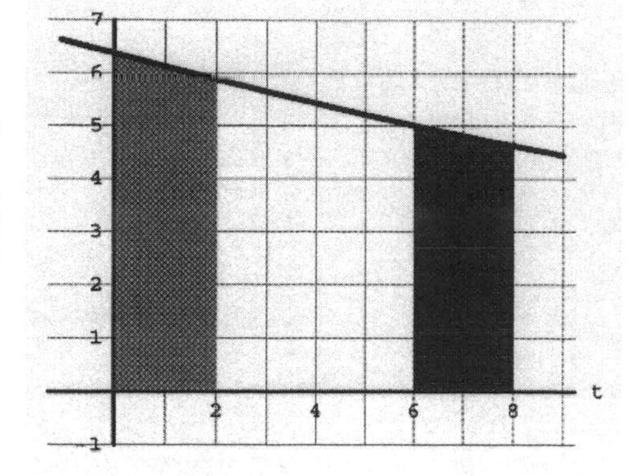

A comparison of the areas between the graph of $f(t) = 6.37e^{-0.04t}$ and the t axis on the interval from t = 0 to 2 hours (the light shaded region) and from t = 6 to 8 hours (the dark solid region).

This information can help managers determine when employees should take breaks so that they can optimize their performance, because they would likely feel more productive when they returned to work.

A definite integral can help heating and cooling companies estimate the amount of costs needed to send power or gas to each household. On any given day, the temperature can be modeled with a sinusoidal function, because temperature increases during the day, decreases at night, and then repeats the cycle throughout the year. For example, suppose the temperature reached a low of 50° Fahrenheit at 2 AM and a high of 90° at 2 PM. If x represents the number of hours that have passed during the day, then the temperature in degrees Fahrenheit, T, can be represented by the equation $T = 20\cos\left[\frac{2\pi(x-14)}{24}\right] + 70$. Suppose that the thermostat in the house is set to 80° so that the air conditioning will turn on once the temperature is greater than or equal to that setting. The amount of energy used for the air conditioner is proportional to the temperature outside. That means that the air conditioner will use more energy to keep the house cool when it is closer to 90° than when it is near 80°. The price to cool the house might be five cents per hour for every degree above 80°. If the temperature were 83° for the entire hour, then the cost to run the air conditioner would be fifteen cents. However, since temperature varies according to a sinusoidal function, the cost per minute would actually change. Therefore, a definite integral bounded by the time when the temperature is above 80° will help predict the cooling costs. The temperature should be 80° at $x = 10$ (10 AM) and $x = 18$ (6 PM), so the cooling costs per day for days like this would be approximately $2.62 based on an evaluation of the expression

$$\$0.05\int_{10}^{18}(20\cos\left[\frac{2\pi(x-14)}{24}\right] + 70 - 80)dx = \$2.62.$$

Notice that the answer is also represented by 0.05 times the area of the curve between $T = 20\cos\left[\frac{2\pi(x-14)}{24}\right] + 70$ and $T = 80$, as shown in the following diagram.

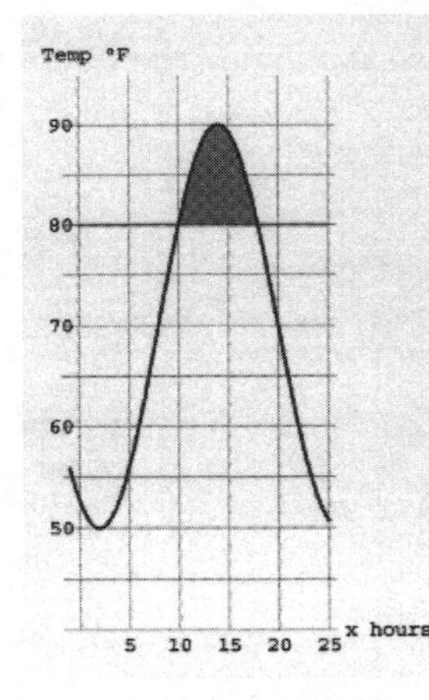

The area between $T = 20\cos\left[\frac{2\pi(x-14)}{24}\right] + 70$ *and T = 80, which is the same as* $\int_{10}^{18}(20\cos\left[\frac{2\pi(x-14)}{24}\right] + 70 - 80)dx$

Integration can be used to help solve differential equations in order to formulate new equations that compare two variables. A *differential equation* is a relationship that describes a pattern for a rate. For example, the differential equation describing the rate of the growth of a rabbit population is proportional to the amount present and would be represented by the equation $\frac{dP}{dt} = kP$, where P is the population, t is the amount of time, and k is a constant of proportionality. If there were 200 rabbits in the population seven months ago, and 500 rabbits in the population right now, then an integral will help you find an equation that relates the population of rabbits to the amount of time that has passed. In this case, solving the differential equation will result in a general equation of $P = 200e^{0.131t}$, where t is the number of months that have passed since the rabbits were originally counted. This information can help farmers understand how their crops will be affected over time and take preventative measures, since they will be able to predict future rabbit populations, assuming that changes will not result in the growth rate due to disease or removal.

The equation $d = 0.5gt^2 + v_ot + d_o$ is commonly used in physics when studying kinematics to describe the vertical position, d, of an object based on the time the object has been in motion, t. Values that are commonly substituted into this equation are $g = -9.8$ meters per second squared to represent the acceleration due to earth's gravity, the initial velocity of the object, v_o, and the initial vertical position of the object, d_o. How was this equation determined? Integration can help explain how this expression is derived.

The acceleration of an object in vertical motion is equal to the constant value, g, neglecting any air resistance. Acceleration is a rate of velocity, v, so $v = \int g dt$. The velocity at $t = 0$ is v_o, so this information and the integral determines the equation $v = gt + v_o$. Velocity is a rate of position, so $d = \int(gt + v_o)dt$. The vertical position at $t = 0$ is d_o, so this information and the integral determine the equation $d = 0.5gt^2 + v_ot + d_o$.

Many volume formulas in geometry can also be proven by integration. In this case, the integral serves as an accumulator of small pieces of volume until the entire solid is formed. For example, the volume, r, of a sphere can be represented by the equation $v = \frac{4}{3}\pi r^3$, where r is the radius of the sphere. This equation can be determined by revolving a semicircle, $y = \sqrt{r^2 - x^2}$, about the x-axis.

A sphere with radius, r, is formed when rotating the semicircle $y = \sqrt{r^2 - x^2}$ *about the x-axis.*

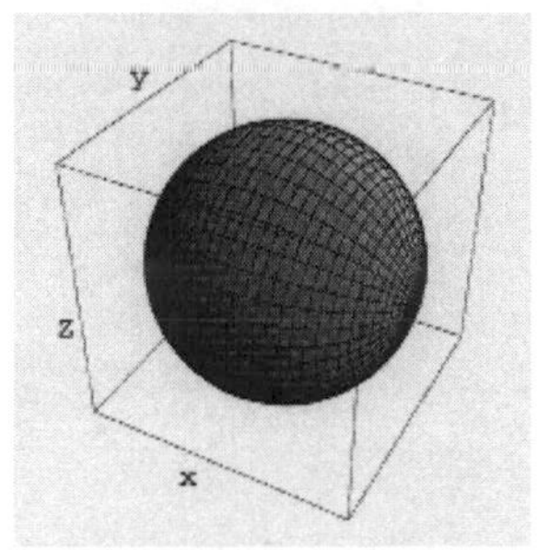

One really thin cross-sectional slice of the sphere can be represented by a cylinder with radius y and thickness Δx, as shown in the left-hand diagram below. The volume of this cylindrical cross section, then, is $v = \pi y^2 \Delta x$. The integral will accumulate the volume of all of these cylinders that stack up against one another from $x = -r$ to $x = r$.

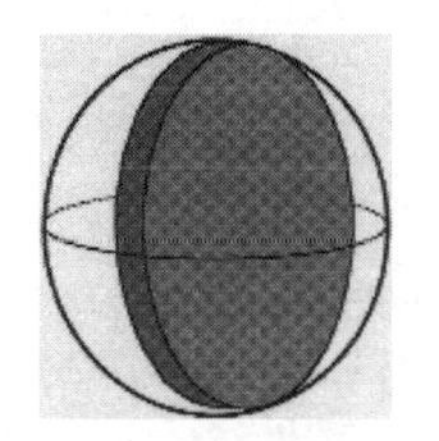

cylindrical slice inscribed in a sphere

cylindrical slices stacked to form a sphere

Cylindrical slices with radius $y = \sqrt{r^2 - x^2}$ *and height* Δx *are stacked together to form the volume of the sphere.*

Therefore the volume of a sphere can be represented by $\pi \int_{-r}^{r} (\sqrt{r^2 - x^2})^2 dx$, which simplifies to $v = \frac{4}{3}\pi r^3$. This formula tells manufacturers how much metal is needed to create certain ball bearings. The formula is also useful for ice cream store owners to determine how many cones they can serve with each container of ice cream, assuming that they can convert cubic centimeter units to gallons. Orange juice manufacturers can use this relationship to estimate the amount of orange juice they will receive from a batch of fresh oranges.

What about predicting the volume needed to juice other fruits that have non-circular curves, such as lemons, apples, and pears? The process would be similar to calculating the volume of a sphere, except that an equation would need to be developed to model the perimeter of the fruit. For example, if the core of a pear is placed along the x-axis, a pencil can trace its perimeter in the first two quadrants. A fourth-degree function can model the curvature of a pear, such as $y = -0.016x^4 - 0.094x^3 - 0.068x^2 - 0.242x + 3.132$, and then rotated around the x-axis to form the solid, as shown below. An integral set up like the volume of a sphere, and bounded by the x-intercepts of the function, will approximate the volume of the pear.

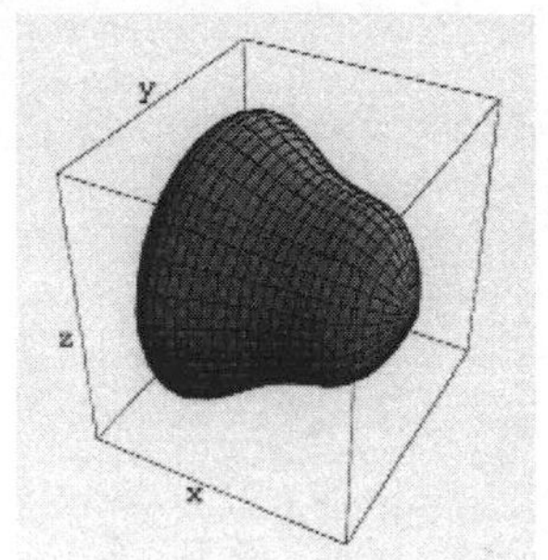

A pear can be constructed by rotating the function $y = -0.016x^4 - 0.094x^3 - 0.068x^2 - 0.242x + 3.132$ *about the x-axis.*

This pear has a volume ≈ 352 milliliters, as a result of evaluating

$$\pi \int_{-6.347}^{2.502} (-0.016x^4 - 0.094x^3 - 0.068x^2 - 0.242x + 3.132)^2 dx.$$

online sources for further exploration

BHS calculus student projects
<http://www.bhs-ms.org/calculus.htm>

The case of the murky mell
<http://www.math.iupui.edu/writing_in_math/murky_well.html>

CO_2 concentrations in a river
<http://www.geom.umn.edu/education/calc-init/integration/>

Flood levels
<http://www.math.bcit.ca/examples/ary_15_6/ary_15_6.htm>

Gavin's calculus projects
<http://www.math.lsa.umich.edu/~glarose/courseinfo/calc/calcprojects.html>

Heating-degree-days
<http://www.nap.edu/html/hs_math/hd.html>

Internet differential equations activities: Current projects
<http://www.sci.wsu.edu/idea/current.html>

Modeling population growth
<http://www.geom.umn.edu/education/calc-init/population/>

Nuclear medicine
<http://www.math.bcit.ca/examples/ary_11_6/ary_11_6.htm>

Petroleum collection
<http://www.math.bcit.ca/examples/ary_13_6/ary_13_6.htm>

Surveying
<http://www.math.bcit.ca/examples/ary_17_6/ary_17_6.htm>

Tunnel Vision, Inc.
<http://panther.bsc.edu/~bspieler/projects/tunnel.html>

INVERSE (MULTIPLICATIVE)

A relationship in the form $y = \frac{k}{x}$, where k is a constant, is called an *inverse function*. Sometimes you will see this relationship written as "y is inversely proportional to x." The graph of this function is a hyperbola, but most real-world applications with inverse functions relate only to nonnegative values in the domain.

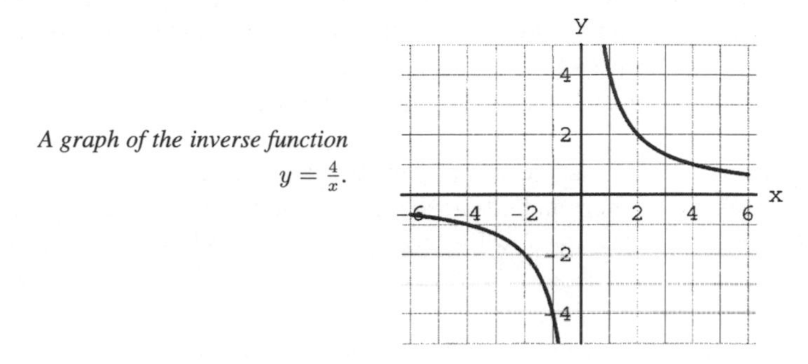

A graph of the inverse function $y = \frac{4}{x}$.

An inverse proportion indicates that the dependent variable decreases as the independent variable increases, or vice versa. In other words, as one factor changes, the other factor will change in the reverse direction. For example, pressure is inversely proportional to the volume of an object. When you dive underwater, the amount of air space in between your ears begins to decrease, causing them to pop, because the pressure gradually increases. If your ears do not pop and release the air inside, then you will feel discomfort or pain, because the pressure becomes too great.

Bottling companies use this same principle in packaging their soft drinks. Air and carbon dioxide are compressed in a small volume of space when you first open a container. The built-up pressure inside the small amount of space will cause the container to fizz or make a sound when it is first opened. After the gasses in the container have been released and part of the bottle is emptied, the pressure on the bottle decreases, since the air volume inside has increased. Thus the bottle does not fizz as much when it is opened later on.

Size is influenced by a combination of surface area and volume. The ratio of surface area to volume is an inverse relationship, because area units are squared and volume units are cubed. The ratio of squared units to cubic units is equal to inverse units. For example, the surface area of a cube with an edge length equal to 2 cm is 24 cm^2 (6 square faces, each with area of 4 cm^2). The volume of this cube is 8 cm^3. The surface-area-to-volume ratio is 3 cm^{-1} (determined by $\frac{24\ \text{cm}^2}{8\ \text{cm}^3}$). Notice that the units are a multiplicative inverse, or reciprocal, of cm.

Since an organism's metabolic rate is affected by this ratio, it can be modeled after an inverse proportion. This means that large animals will typically have slower metabolisms than smaller animals, because the ratio of surface area to volume will decrease for larger volumes. Conversely, smaller animals will have higher metabolic rates than larger animals, because this ratio increases for smaller volumes. Therefore rodents and dogs are much more likely to lose heat from their bodies than bears and elephants, because they have less heat stored inside. As a result, smaller animals need to be more active to maintain appropriate heat levels within their bodies, causing their metabolism to remain at high levels. Animals and plants have naturally developed parts of their body to help expand their surface area without adding considerable volume so that they can increase their metabolic rate. For example, trees develop leaves from branches, and humans use capillaries to extend their circulatory system. Microvilli, the lining of the small intestine, is an example of a large surface area in the human body with little volume, because it stretches to lengths of over seven meters long!

Fuel consumption as a function of gas mileage is an inverse relationship. As automobiles increase their fuel efficiency, or the number of miles per gallon they attain while driving, then the gasoline consumers will purchase less fuel. Smaller compact cars typically obtain better gas mileage, because there is less mass to move when compared to less-fuel-efficient cars such as vans, trucks, and sport-utility vehicles. If Americans drive approximately 10^{12} miles each year, then the fuel consumption of the United States each year can be represented by the function $g = \frac{10^{12}}{m}$, where m is the average gas mileage of the cars that year.

Production rates also form inverse relationships. The time it takes to complete a task is inversely proportional to the rate at which an item is produced or performed. For example, a grocery store clerk needs to staple price stickers on 5,000 cans. The amount of time needed for the job, t, is dependent on his productivity rate, r, according to the function $t = \frac{5000}{r}$. If he works at a faster rate, then the job will take less time to complete.

Some people claim that "Murphy's law" can be described as an inversely proportional relationship. This law maintains that anything can go wrong at the worst possible moment. For example, when constructing a stage, Murphy's law might strike if the most vital tool to complete the job is missing. Another example is staying up all night to complete a term paper, only to realize that your disk has gone bad or your printer is out of ink. If this law were described as an inverse function, then the availability of an item or luck is inversely proportional to its importance. Thus as an event or object becomes more important, Murphy's law can strike, indicating that it will likely not occur or be available. Conversely, the object or phenomenon will more likely occur or become available when it is less needed.

online sources for further exploration

Best practices in network security
<http://www.silcom.com/~aludwig/Physics/Main/Image_analysis.html>

Boyle's law
<http://library.thinkquest.org/12596/boyles.html>

Ears, altitude and airplane travel
<http://www.entnet.org/altitude.html>

Gear ratios
<http://www.meceng.uct.ac.za/~mec104w/projects/legogears/legogears.html>
<http://weirdrichard.com/gears.htm>

Investigating direct and inverse variation with the telescope
<http://jwilson.coe.uga.edu/emt669/Student.Folders/Jeon.Kyungsoon/IU/rational2/Telescope.html>

Murphy's law
<http://www.peacockfamily.co.nz/murphys.html>
<http://fun.pinknet.cz/wise/m_apl.htm>

Weight and distance on a lever
<http://www.indiana.edu/~atmat/units/ratio/ratio_t7.htm>
<http://collections.ic.gc.ca/science_world/english/exhibits/leverarm/index.html>
<http://www.pbs.org/wgbh/nova/teachersguide/lostempires/lostempires_sp3.html>

▲ ▼ ▲

INVERSE FUNCTION

An *inverse* is a process, procedure, or operation that is reversed. For example, the inverse of walking up the stairs is walking down the stairs. The inverse of putting on your socks and then your shoes in the morning is taking them off at night. When you are given driving directions to a friend's house, you have to use the inverse of the original directions to find your way home, because all of the directions will need to be reversed, where left turns will become right turns, and vice versa.

Two functions, $f(x)$ and $g(x)$, are inverses if their composites are equal to the independent variable. Symbolically, this is written $f(g(x)) = x$ or $g(f(x)) = x$. Also, the coordinates on inverse functions are reversed. If $f(x)$ and $g(x)$ are inverses, and $f(x)$ contains the point (4,7), then $g(x)$ contains the point (7,4). So one way to model an inverse of a function is to reverse the coordinates. For example, the exchange rate when traveling from the United States to Australia might be represented by the function $a = 1.90u$, where u is the number of U.S. dollars, and a is the number of Australian dollars. This means that 1.90 Australian dollars is equivalent to 1 U.S. dollar. In this equation, the coordinates are represented by the ordered pair (u, a). The inverse of this relationship would be to describe the exchange rate when traveling from Australia to the United States. Therefore the coordinates would be reversed, or (a, u). In order to represent this equation as a function that indicates the exchange rate, the equation $a = 1.90u$ needs to be rewritten as a function in terms of a. This can be done by dividing both sides of the equation by 1.90 to obtain $\frac{1a}{1.90} = \frac{1.90u}{1.90}$, which simplifies to approximately $0.53a = u$. This means that the exchange rate on the return to the United States is about $0.53 (U.S.) for every Australian dollar. You can verify that these two functions are inverses by finding their composite $u(a(u))$, which equals $0.53(1.90u)$ and simplifies to equal u.

Inverses are also used to decode secret messages. If an encrypting pattern is used to change the letters in a sentence, then a decrypting pattern is needed to place the letters back in their normal positions. For example, suppose an encryption function of $e(w) = 2w + 3$ is applied to each letter in a word, where letters had corresponding numbers (e.g., a = 1, b = 2, c = 3, . . . z = 26). The word "math" would first translate to a numerical expression, 13 1 20 8, and then be transformed using the function $e(w) = 2w + 3$, where w represents the original number, and e represents the coded number. So the letter "m," equivalent to 13, would transform to $e = 2(13) + 3 = 29$. After transforming 1, 20, and 8, the final coded expression would be 29 5 43 19. If you receive the secret transmission of an encoded expression 29 5 43 19, you will need to decode it using the inverse function, $w(e)$. One way to find this inverse is to solve for w in the equation $e = 2w + 3$ by subtracting 3 from both sides and then dividing by 2. Thus $w = \frac{e-3}{2}$ is the inverse operation that will decode the expression. The first number, 29, would be converted to $w = \frac{29-3}{2} = 13$, which translates to the letter "m" in the alphabet. If you apply the inverse function to the remaining numbers 5 43 19, you will obtain the word "math" again. Encrypted codes that deal with confidential information, such as credit card numbers and highly classified material, are far more complex than this function. However, the decryption of obscure codes is often performed by computers that use a program that relies on the process of an inverse operation! (For a more secure code, see ***Matrices***.)

online sources for further exploration

Cryptology
<http://www.ssh.fi/tech/crypto/intro.cfm>
<http://www.jcoffman.com/Algebra2/ch4_5.htm>
<http://www.sans.org/infosecFAQ/encryption/mathematics.htm>

Inverse problems in the earth sciences
<http://ees-www.lanl.gov/EES5/inverse_prob.html>

Universal currency converter
<http://www.xe.net/ucc/>
<http://www.wildnetafrica.com/currencyframe.html>

INVERSE SQUARE FUNCTION

The *inverse square law* explains why sound drops off so quickly as you move away from a source of noise, why porch lights do a good job of illuminating the front of a house but not the street in front of the house, and why the forest reverts to darkness as you move away from a campfire. Light and sound emerging from single sources can be viewed as increasing spheres whose area increases as the square of the distance from the source. As a result, the proportion of sound or light reaching a specific unit of area, such as a square meter, varies inversely as the square of the distance from the source. The standard equation of an *inverse square function* is $y = \frac{k}{x^2}$, where k is a constant of proportionality.

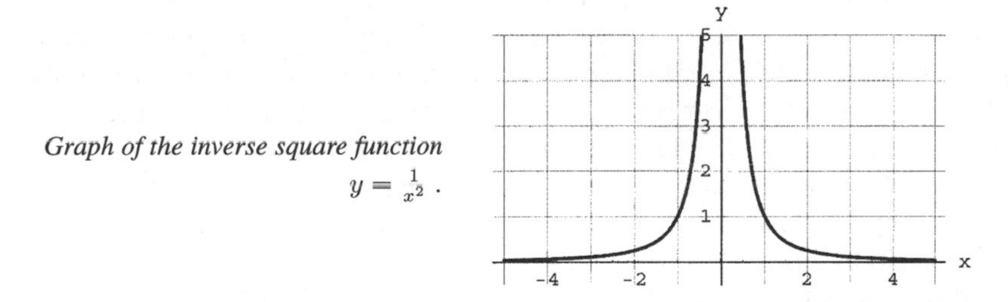

Graph of the inverse square function $y = \frac{1}{x^2}$.

Light emerges from a source in all directions. The drawing below illustrates the distribution of light from a 40-watt light bulb. We will use wattage as the measure of pointance and illuminance to show how the inverse square works. Imagine a sphere of radius r containing the light bulb. Forty watts fall on the interior surface of the sphere. The surface area of the sphere is given by $SA = 4\pi r^2$ square meters. The energy falling on 1 square meter is therefore 40 watts divided by $4\pi r^2$. In the drawing, this energy is called "L." If you go out twice as far, the same energy is distributed over 4 square meters. At three times the distance, the energy is distributed over 9 square meters.

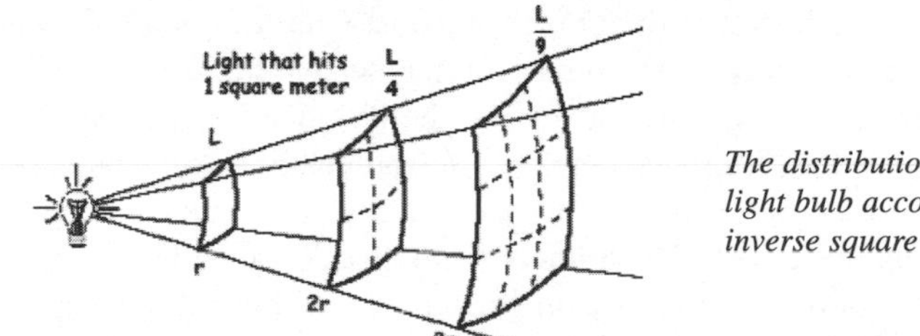

The distribution of light from a light bulb according to the inverse square law.

A standard formula for light intensity is $E = \frac{I}{r^2}$, where E is called the *illuminance*, and I is called the *pointance*. Illuminance is measured in a variety of units such as lux and footcandles. Pointance is a measure of the intensity at the source of the light. As you can see, this is a direct statement of the inverse square law.

The inverse square law provides information about the likelihood of other planets having life forms similar to those on earth. Imagine the light that hits your neighborhood at noon on a hot summer day. Call that amount of light L. Mars is about one-and-a-half times the distance from the sun as the earth, so if your neighborhood were on Mars, it would receive $\frac{L}{1.5^2}$ or 44 percent of the light it receives on earth. That might be enough to sustain life. The planet Neptune is about thirty times as far from sun as the earth. If your neighborhood were on Neptune, it would receive $\frac{L}{30^2}$, or about 0.1 percent of the light it receives on earth. This wouldn't be enough to support life as we know it.

The distribution of sound follows the same rule. Just replace the light bulb in the first illustration by an actor in an auditorium. In an auditorium, the sound drop due to the inverse square law is usually unacceptable. It would mean that a person standing in front of the first row of seats, who might be audible to people in the tenth row, would be barely audible to people in the twentieth row. The audibility of the speaker (about 70 decibels) to listeners in the first row would drop to 50 decibels (a soft sound) ten rows behind. Acoustical engineers design reverberation into auditoriums to focus the sound and overcome the inverse square law. They place hard surfaces at the back of the stage and on the ceiling and walls so that sound that would ordinarily dissipate would bounce back and add to the intensity of that being heard by the audience.

Gravity is an example of a force that follows the inverse square law. How much lighter will a 160-pound astronaut feel if he or she is in a spaceship 12,000 miles above the earth? The radius of the earth is about 4,000 miles, so the astronaut is 16,000 miles from the center of the earth, or about four times the distance of a person measuring weight on the surface of the earth. By the inverse square law, the astronaut would feel as though his or her weight were $\frac{160}{4^2} = 10$ pounds, even though the mass of the astronaut remains unchanged.

Electric force acting on a point charge, q_1, in the presence of another point charge, q_2, is given by *Coulomb's law*, $F = \frac{kq_1q_2}{r^2} = \frac{q_1q_2}{4\pi\epsilon_0 r^2}$, where ϵ_0 is the constant for the permittivity of free space. This law is an outcome of the inverse square law. It is named in honor of the French scientist Charles Coulomb, who established it in 1777 after studying the forces on magnetized needles.

The inverse square law means that increasing the distance from a source of nuclear radiation may be the difference between life and death. Accidental exposure to radiation that produces 600 rems (a measure of radiation impact on living tissue) is almost certain to cause death within two months. A person who is twice as far away will absorb 600/4 = 125 rems, an amount that will result in a significant, but temporary, reduction in blood platelets and white blood cells. If the radiation distribution followed an inverse law, rather than an inverse square law, then a person twice as far away as the one receiving the fatal dose would get 600/2 = 300 rems. This dose causes severe blood damage, nausea, hair loss, hemorrhage, and death in many cases. Because radiation follows the inverse square law, being twice the distance from a fatal dose may mean illness rather than death.

The inverse square law comes up in court cases. The lawyer faced the medical examiner and asked suddenly, "The body wasn't found in the bedroom. How can you say that the fatal shots were made there?" The examiner replied, "Because we found blood spatters on the wall in the bedroom. Measuring the number of splatters in a square meter, we used the inverse square law to determine that a high-velocity bullet passed through the victim 1.7 meters from that wall. Analysis of droplet angles confirmed our estimate."

online sources for further exploration

Summary from the Jet Propulsion Laboratory
<http://www.star.le.ac.uk/edu/solar/edu/invsquar.html>
<http://www.solarviews.com/span/edu/invsquar.htm>
<http://www.jpl.nasa.gov/basics/bsf6-1.html>

Physics examples and lessons
<http://hyperphysics.phy-astr.gsu.edu/hbase/forces/isq.html>
<http://www.physicsclassroom.com/Class/circles/U6L3b.html>

Universal law of gravitation
<http://csep10.phys.utk.edu/astr161/lect/history/newtongrav.html>

Simple experiments
<http://www.exploratorium.edu/snacks/inverse_square_law.html>
<http://www.public.iastate.edu/~javapgmr/homepage.html>
<http://www.howstuffworks.com/question441.htm>

Astrophysics
<http://jersey.uoregon.edu/vlab/InverseSquare/index.html>
<http://www.solarviews.com/span/edu/invsquar.htm>
<http://www.star.le.ac.uk/edu/solar/edu/invsquar.html>

Basic notions of celestial mechanics
<http://www.rafed.net/arc/arabic/research/mmar/bncm/bncm1.htm>

Darkness outside of a campfire
<http://www.exploratorium.edu/snacks/inverse_square_law.html>

Electromagnetic radiation
<http://www.jpl.nasa.gov/basics/bsf6-1.html>

How light works
<http://www.howstuffworks.com/question441.htm>

Simulated ocean dive
<http://illuminations.nctm.org/imath/912/Light/light2.html>

LINEAR FUNCTIONS

A *linear function* is a function that has a constant change in the dependent variable for every change in the independent variable. For example, the value of the dependent variable y in the linear function $y = 5x - 2$ will always increase by five units for every increase in one unit of the independent variable, x. This means that the ratio of these changes, called the *slope*, is also constant. For example, the previous comparison is the same as saying that there will be a change of fifteen units in the dependent variable for every change in three units of the independent variable, since this ratio simplifies to 5. Every linear function can be written in the slope-intercept form, $y = mx + b$, where m is the slope of the line, and b is in the y-intercept.

Realistic situations use linear functions to make predictions or draw comparisons that involve constant change. For example, the cost of gasoline is linearly related to the number of gallons purchased. For every one gallon of gas purchased, the price will increase approximately \$1.40. The fact that the gas price per gallon does not change as gas is pumped allows someone to use a linear function to predict the amount of money needed to fill the tank. In this situation, the function $c = 1.40g$ would relate the cost in c dollars to g gallons purchased. If an automobile has a twelve-gallon tank, then the cost to fill the tank would be $c = 1.40(12) = \$16.80$. In addition, the linear equation is useful when the individual purchasing gasoline would like to know how much gasoline he or she would obtain with the \$10 available in his or her pocket. In this case, 10 would be substituted for the variable c, and solving the equation would show that approximately 7.14 gallons could be purchased, slightly more than half a tank in most cars.

Linear functions are useful in estimating the amount of time it will take to complete a road trip. Assuming that traffic conditions are good and the driver is traveling at a constant speed on a highway, the linear equation $d = rt$ (distance equals rate times time) can be used to predict the total distance traveled or time needed to complete the trip. For example, suppose that a family is traveling on vacation by automobile. The family members study a map to determine the distance between the cities, estimate a highway speed or rate of 65 miles per hour, and then solve the linear equation $d = 65t$ to estimate the length of their trip. An awareness of the time needed for the trip would likely help the family plan a time of departure and times for rest stops.

Banking institutions determine the amount of simple interest accumulated on an account by using the linear equation $I = Prt$, where I is the amount of interest, P is the initial principal, r is the interest rate, and t is the time in years in which the interest has been accumulating. For example, a \$1,000 loan with 8 percent simple interest uses the function $I = 1000(0.08)t$, or simplified to $I = 80t$, to predict the amount of interest over a specific time period. Once the principal and interest rates have been determined, the function is linear, since the amount of interest increases at a constant rate over time. Over five years, there will be $I = 80(5) = \$400$ net payment in interest.

Circuits rely on linear relationships in order to operate electrical equipment. The voltage V, current I, and resistance R are related with the equation $V = IR$. A power supply has voltage to create a stream of current through electrical wires. The current in a circuit is typically held constant, such as at 72 Hz, so that there is a constant stream of electricity. In this case, the linear relationship $V = 72R$ would help a manufacturer determine the amount of resistance needed in a power supply so that an electrical object can operate correctly. Resistors are small devices that block or slow down the current so that an object does not receive too much power. For example, if the resistance in a light circuit is too low, then the bulb would receive an overload of power and be destroyed. If the resistance is too high, then there will not be enough power reaching the bulb in order for it to light well. These problems can arise with some appliances when they are moved to different countries, because the electrical circuits may run with different current levels. Consequently, appliances may have different types of resistors so that they can accommodate to the corresponding current levels in a circuit.

In a business setting, a linear function could be used to relate the total costs needed to sell a product in terms of the number of products produced. For example, suppose a bakery created cookies at a raw material expense of \$0.25 per cookie. Suppose production costs for equipment are an extra \$500. In this case, the linear function $t = 0.25c + 500$ will represent the total cost, t, needed to produce c cookies. In general, if a function is modeled by a linear relationship, then the rate (\$0.25 per cookie) will be the slope, and the starting amount (\$500 equipment expense) will be the y-intercept of the equation. This information is useful to the owner, because he or she will be able to predict the average cost of producing cookies, start-up expenses included, or the amount of cookies that can be produced based on a fixed budget.

Unit conversions are often linearly related. For example, the United States uses a different temperature scale (Fahrenheit) than most of the rest of the world (Celsius). If an individual from the United States travels to Spain, then a temperature of 30° Celsius would feel considerably different from a temperature of 30° Fahrenheit. The equation that converts the two variables can be determined by using the freezing and boiling points of water. Water freezes at 0° Celsius and 32° Fahrenheit; water boils at 100° Celsius and 212° Fahrenheit. These two pieces of information represent two ordered pairs on a line, (0,32) and (100,212). Since two points are sufficient information to determine the equation of a line, the slope formula and y-intercept will lead to the equation $F = \frac{9}{5}C + 32$, where F is the temperature in Fahrenheit, and C is the temperature in Celsius. This means that a report of 30° weather in Spain suggests that the day could be spent at the beach, while in the United States a report of 30° weather means that you might be having snow!

Linear functions can be used to form relationships between data that are found in natural events and places. For example, there is a strong relationship between the winning time of the men's Olympic 100-meter dash and the year in which it occurs. The graph that follows shows that a line can be drawn to approximate the relationship between these two variables. Notice that all of the data values do not fall on the line, but instead cluster around it. It is possible for points to be away from the line, especially during years of unusually exceptional performance. The *correlation coefficient*, r, is a measure of the strength of the linear relationship. The relationship is stronger as the absolute value of the correlation coefficient approaches the value of 1. If the correlation coefficient is closer to 0, then a linear relationship does not likely exist. In the 100-meter dash situation, the absolute value of r is equal to 0.88, indicating that the line is a pretty good model for the data.

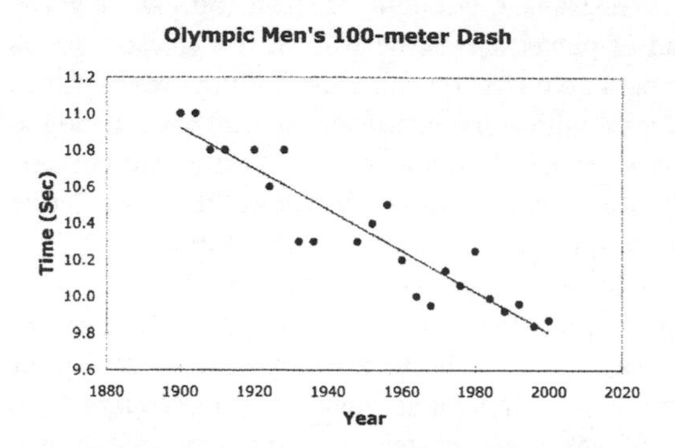

The line that models the pattern of the winning times in the men's Olympic 100-meter dash is Predicted Time = -0.01119Year + 32.185.

The linear equation acts as an approximate prediction of the relationship between time and year. This predicted pattern is much more reliable within the range of data, so the variables may not have the same relationship for future Olympics. After all, the line should eventually level off, because the runners will never be able to run a time equal to zero! Therefore, this line is most useful to make predictions between 1900 and 2000, such as estimating the winning times when the Olympics did not occur or when participation was reduced (often due to world conflicts). For example, there is no time for 1944 because the Olympics were suspended during World War II. The time that might have been achieved in the 1944 Olympics could be estimated using the linear model *Predicted Time* = –0.01119*Year* + 32.185 by substituting 1944 for *Year*. That gives a predicted winning time of 10.43 seconds. Linear relationships are also common with winning times and championship performances in many other Olympic events.

Forensic scientists use linear functions to predict the height of a person based on the length of his arm or leg bones. This information can be useful in identifying missing people and tracing evolutionary patterns in human growth over time. When a complete skeleton cannot be found, then the height of the deceased person can be predicted by identifying the person's sex and finding the length of his or her femur, tibia, humerus, or radius. For example, the height h in centimeters of a male can be estimated by the linear equation $h = 69.089 + 2.238f$, where f is the length of the femur bone in centimeters. In addition, the linear equation $s = -0.06(a - 30)$ or $s = -0.06a + 1.80$ is the amount of shrinkage s for individuals of age a greater than 30 that needs to be accounted for in the height of a deceased person. For example, if the person had an estimated age of 60 at death, then $-0.06(60) + 1.80 = -1.80$ cm would be included in the height prediction.

Ever feel cold in an airplane? The outside temperature decreases linearly with an increase in altitude. The equation $t = -0.0066a + 15$ has been described as a linear model that compares the temperature t (°C) with the altitude a (meters) when the ground temperature is 15°. Recognizing this relationship helps engineers design heating and cooling systems on the airplanes so that metal alloys can adapt to the changes in temperature and passengers obtain reasonable air temperatures inside the plane. The linear equation also helps pilots understand the limitations as to how high they can fly, because not only are there changes in air pressure, but the temperature decreases by 66° for every 10,000 meters of altitude. You would not want the wind blowing in your face at high altitudes!

The apparel industry uses linear functions when manufacturing dresses. Dress sizes often reflect a general relationship among a woman's bust, waist, and hips. The table below shows the relationship among these measurements.

size *s*	6	8	10	12	14	16	18	20
bust *b*	30.5	31.5	32.5	34	36	38	40	42
waist *w*	23	24	25	25.5	28	30	32	34
hips *h*	32.5	33.5	34.5	36	38	40	42	44

A comparison of women's dress sizes according to bust, waist, and hips (dimensions in inches).

The data can be generalized into a few linear relationships. The designer estimates that a woman's dress size is $s \approx 1.2w - 20$. Furthermore, the other measurements can be approximated with the linear relationships, $b \approx 1.1w + 5$ and $h = b + 2$, making the equations a reasonable predictor of all measurements and sizes, including those that are not listed. These relationships allow manufacturers to mass produce dresses and provide women with a general reference point for clothing sizes. However, since women have different body types, dresses are sometimes altered or designed in different ways to accommodate the needs of a variety of consumers.

Linear models have also been used within political arenas to argue for legislation. For example, the state of Florida had been confronted with the problem of powerboat speeds along its waterways that affect the survival of the manatees, which are very large but docile creatures that live in shallow water. Because they swim on the surface and near shore, many manatees have been killed or injured by blades of powerboats. Lobbyists concerned about the death of the manatees were able to show a strong linear relationship between the number of their deaths and the number of powerboat registrations for the years 1977 to 1990. The equation is *Killed* = 0.125 *Powerboats* − 41.430, where *Powerboats* is the number of

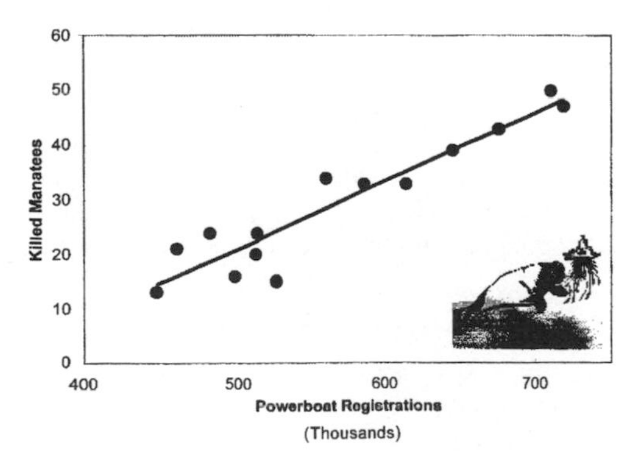

Number of manatees killed in Florida related to the number of powerboat registrations for the years 1977 to 1990.

thousands of powerboats registered in Florida. The slope indicates that for every thousand more powerboats registered, 0.125 more manatees are killed. In other words, for every 10,000 more powerboats, the slope predicts 1.25 more manatees are killed.

The data and linear function created a compelling argument that the manatees were at risk of being endangered in a short period of time, unless action was taken to reduce the number of powerboats and to slow down their speed in shallow waters. As a result, the Florida legislature has made it more difficult and expensive to acquire a powerboat license. It increased the number of game and fish officers in manatee areas so that "no wake" rules would be strictly enforced.

online sources for further exploration

Battery depletion and piecewise linear graphing
<http://daniel.calpoly.edu/~dfrc/Robin/Pathfinder/Battery/batt.html>

Cassette tape project
<http://ericir.syr.edu/Virtual/Lessons/Mathematics/Functions/FUN0001.html>

The diet problem
<http://www-fp.mcs.anl.gov/otc/Guide/CaseStudies/diet/>

Discovering the linear relationship between Celsius and Fahrenheit
<http://daniel.calpoly.edu/~dfrc/Robin/Celsius/celsius.html>

Environmental health
<http://www.math.bcit.ca/examples/ary_8_2/ary_8_2.htm>

Linear regression with human movements
<http://exploringdata.cqu.edu.au/lin_reg.htm>

Nuclear medicine
<http://www.math.bcit.ca/examples/ary_11_2/ary_11_2.htm>

Olympic statistics
<http://www.swishweb.com/Sports_and_Games/Olympics/>

Plotting temperature and altitude
<http://daniel.calpoly.edu/~dfrc/Robin/Pathfinder/Temp/temp.html>

Property lines
<http://www.math.bcit.ca/examples/ary_17_8/ary_17_8.htm>

Size effects on airplane lift
<http://www.grc.nasa.gov/WWW/K-12/airplane/size.html>

Timing traffic lights
<http://www.nap.edu/html/hs_math/tl.html>

Voltage circuit simulator
<http://jersey.uoregon.edu/vlab/Voltage/index.html>
<http://java.sun.com/applets/archive/beta/Voltage/index.html>

Your weight on other worlds
<http://www.exploratorium.edu/ronh/weight/index.html>

LOGARITHMS

Logarithms are exponents, so they are used to reduce very large values into smaller, more manageable numbers. It is easier to refer to the number 13.4 than the number 25,118,900,000,000, which is approximately equal to $10^{13.4}$. A number x is said to be the base b logarithm of a number y, if $y = b^x$. The corresponding logarithmic equation is $x = \log_b y$. Base-10 logarithms are used to change numbers to powers of 10. For example, $500 \approx 10^{2.69897}$, so 2.69897 is said to be the base-10 logarithm of 500. This is commonly written as $\log 500 \approx 2.69897$. The decimal part ".69897" is called the *mantissa*, and the integer part "2" is called the *characteristic*. Until inexpensive calculators made it easy to do multiplication, division, and roots, scientists and engineers used base-10 logarithms to simplify computations by changing multiplication of numbers into addition of exponents, and division of numbers into subtraction of exponents. Up until twenty years ago, the main computational device for high school students in advanced math and sciences was based on logarithmic scales—the slide rule.

Other common bases for logarithms are the numbers e and 2. The number $e \approx 2.718281828459$. It can be developed from the compound-interest formula as the limit of $(1 + 1/n)^n$ as n increases without bound. The base e is used in exponential expressions that evaluate continuously compounded interest. Logarithms to the base e are typically written with the abbreviation ln, called a *natural logarithm*. $\ln(500) \approx 6.21461$, because $500 \approx e^{6.21461}$. Mathematical functions using e and ln simplify computations with rates and areas that result from situations in physics, biology, medicine, and finance. Hence e and natural logarithms are often used in the statement of rules and properties in these fields. Base-2 logarithms emerge from the study of computer algorithms. Computers are based on on-off switches, so using base-2 logarithms provides a natural connection with machine operations.

Logarithmic scales are used in newspapers, households, and automobiles as well as in scientific research. How loud is a rock concert? Noise is measured in decibels, a logarithmic scale that is easier to use than the sound-energy measurement of watts per square meter. A decibel is one-tenth of a bel, a unit named after Alexander Graham Bell (1847–1922), inventor of the telephone. A soft whisper is 30 decibels. Normal conversation is at 60 decibels. If you are close to the stage at a rock concert, you hear music at 120 decibels. If you are so close that the music hurts your ears, the amplifiers are at 130 decibels. Because the decibel scale is logarithmic, changes along the scale are not linear. When the rock music moves from very loud (120 decibels) to painful (130 decibels), your ears are receiving 10 *times* as much sound energy. The difference of 70 decibels between normal conversation (60 decibels) and pain (130 decibels) represents 10^7 more watts per square meter of sound energy.

People's perceptions of changes in sound intensity are more aligned to the decibel scale rather than the actual changes in energy level. The same goes for the perception of light. The brightness of stars was first put on a quantitative scale by the Greek astronomer Hipparchus at around 130 BC. He arranged the visible stars in order of apparent brightness on a scale that ran from 1 to 6 magni-

sound intensity (watts per square meter)		*relative intensity (decibels)*
10^{3}	Military rifle	150
10^{2}	Jet plane (30 meters away)	140
10^{1}	Pain level	130
10^{0}	Amplified rock music	120
10^{-1}	Power tools	110
10^{-2}	Noisy kitchen	100
10^{-3}	Heavy traffic	90
10^{-4}	Traffic noise in a small car	80
10^{-5}	Vacuum cleaner	70
10^{-6}	Normal conversation	60
10^{-7}	Average home	50
10^{-8}	Quiet conversation	40
10^{-9}	Soft whisper	30
10^{-10}	Quiet living room	20
10^{-11}	Quiet recording studio	10
10^{-12}	Barely audible	0

Decibel levels of common noises.

tudes, with stars ranked "1" as the brightest. Astronomers using powerful telescopes have increased this star-magnitude scale to 29. Analysis of the quantity of light that reaches the viewer indicates that the star-magnitude scale is logarithmic. In the nineteenth century, the scale was standardized so that a difference of 5 magnitudes corresponds to 100 times greater light intensity.

Acidity or alkalinity of a substance is measured on the logarithmic scale pH = $-\log(H^+)$, where H^+ is the concentration of hydrogen ions in moles per liter of the substance. These pH units provide a more compact scale than moles per liter. The scale ranges from 0 to 14, with 7 representing a neutral substance (water). Higher pHs indicate alkalinity, and lower indicate acidic substances. Few plants will survive in soils more acidic than pH = 4 (the acidity of lemon juice) or more alkaline than pH = 8 (baking soda). Battery acid (pH 1) and lye (pH 13) will burn your skin. Litmus papers turn different colors depending on the pH of the substance. A change in color that represents 2 levels of pH will represent a difference of 100 times the concentration of H^+ ions.

The Richter scale is a measure of the strength of earthquakes. An earthquake with a Richter scale value of 4 feels like vibration from a passing train. A scale value of 7 indicates an earthquake that produces ground cracks and causes houses to collapse. Because the scale is logarithmic, the difference in energy from the earthquake waves is 10^{7-4} = 1,000. An earthquake measured as a 7 on the Richter scale is 1000 times more powerful than an earthquake measured at 4.

Logarithms can provide an expression of relations that are inverses of exponential situations. A battery charges at a rate that depends on how close it is to being fully charged. It charges fastest when it is most discharged. The charge C at any instant t is modeled by the formula $C = M(1 - e^{-kt})$, where M is the maximum charge that the battery can hold, and k is a constant that depends on the battery and charger. The formula that gives the time required to charge a battery uses the natural logarithm function ln: $t = -\frac{1}{k}\ln(1 - \frac{C}{M})$.

Logarithms appear in a wide range of industrial and technological applications. The Haugh unit is a measure of egg quality that uses base-10 logarithms. The logarithms of the sizes of two organs of an animal are related in an *allometric equation*. Economists use logarithmic derivatives to compare price changes of different items. The effective steam pressure in a cylinder is $p = \frac{P(1+\ln(R))}{R}$, where P is the initial absolute pressure, and R is the ratio of expansion. The number of turns of a rope or pulley about a large cylinder that would be needed to keep the rope from slipping is found with a formula that uses logarithms of tension ratios. Electrical engineers use Bode plots, a form of logarithmic graphing, to determine voltage gains for active or passive filters. Statisticians use logarithms to linearize data that appear to lie in certain curvilinear patterns.

online sources for further exploration

Investigate pH
<http://www.miamisci.org/ph/>
<http://ga.water.usgs.gov/edu/phdiagram.html>
<http://www.chem.tamu.edu/class/fyp/mathrev/mr-log.html>

e as the base of natural logarithms at the MathSoft site
<http://www.mathsoft.com:80/asolve/constant/e/e.html>

CoolMath's table of decibel levels at
<http://www.coolmath.com/decibels1.htm>

National earthquake information center
<http://wwwneic.cr.usgs.gov/neis/eqlists/eqstats.html>

An excellent list of applications in technical areas is at British Columbia Institute of Technology
<http://www.math.bcit.ca/examples/table.htm>

Sonic booms and logarithms
<http://daniel.calpoly.edu/~dfrc/Robin/Sonic/sonic.html>

Sound pressure levels and intensity
<http://www.coolmath.com/decibels1.htm>
<http://www.math.bcit.ca/examples/ary_12_4/ary_12_4.htm>

Belt friction
<http://www.math.bcit.ca/examples/ary_9_4/ary_9_4.htm>

Bode plots
<http://www.math.bcit.ca/examples/ary_1_4/ary_1_4.htm>

Modeling exponential decay using logarithms (finding half-life)
<http://math.usask.ca/readin/examples/expdeceg.html>

Further examples on logarithms
<http://www.math.bcit.ca/examples/table.htm>

▲▼▲

LOGISTIC FUNCTIONS

Logistic functions predict proportions or probabilities. They are used to determine proportions of successes in "yes–no" situations from underlying factors. They can be used to predict the proportions of students admitted to a university from different SAT-score intervals; the probability of getting an item right on a test depending on underlying knowledge; the probability that a patient with certain symptoms will die or live; the proportions of nerves in the brain that will fire in the presence of different concentrations of stimulating chemicals; the spread of rumors; and the proportion of consumers that will switch brands or stay with their current one when presented with different saturations of advertising.

A logistic function takes the form $y = \frac{1}{\frac{1}{m} + b_0 b_1^x}$, where m is the maximum value of the dependent variable (in most cases, this will be 1.00). The values b_0 and b_1 are very similar to the numbers used in exponential growth models. The illustration below shows the shape of a logistic function. The scatterplot in it shows the percent of applications for admission to a large state university that resulted in acceptances of the candidates. The groupings of students on the x-axis are by SAT verbal score. The dot at 700 indicates that 95 percent of the applicants who had SAT verbal scores at 700 (that is, in the range of 680–720) were accepted. However, only 9 percent of the students at 400 (in the range of 380–420) were accepted. The equation for the logistic curve that models the data is $A = \frac{1}{1+9128(0.983)^{\text{SAT}}}$, where A is the proportion accepted at an SAT score level.

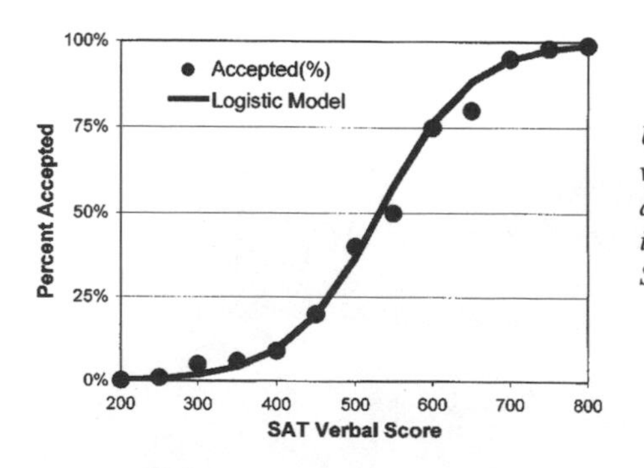

University admission rates for students with different SAT verbal scores. The acceptance rate of high school students into a certain college based on their SAT verbal score.

If you cover up the right side of the curve (SAT verbal scores greater than 550), the remaining curve looks like an exponential curve.

Consider the spread of rumors. Suppose that every hour a person who hears a rumor passes it on to four other people. During the early life of the rumor the equation that represents the spread of the rumor at each hour would be $N = 4^t$, where N is the number of people hearing the rumor at t hours. The exponential growth equation would require 65,536 new listeners at the eighth hour. But what if the rumor starts with a student in a 1,000-student high school overhearing the principal saying, "We are going to dismiss school early today"? If every student passing on the rumor could find someone who had not heard it, then the rumor would pass through the entire student body before five hours were up. However, after four hours, people spreading the rumor will be telling it to students who already know. This means that the rate at which new listeners receive the rumor has to decrease as the day goes on. People who learn about the rumor later in the day are not likely to find anybody who hasn't heard it. A logistic equation that models the spread of this rumor is $N = \frac{1}{\frac{1}{1000}+0.25^t}$, where N is the number of students in the high school who have heard the rumor, and t is the number of hours since the rumor started. This model would predict that half the student body would have heard the rumor by the fifth hour.

Studies of diseases indicate that the early stages of an epidemic appear to show an exponential growth in infected cases, but after a while the number of people infected by the disease does not increase very rapidly. Like the spread of rumors, diseases cannot be easily spread to new victims after much of the population has encountered it. Logistic models describe the number of people infected by a new disease if the entire population is susceptible to it, if the duration of the disease is long so that no cures occur during the time period under study, if all infected individuals are contagious and circulate freely among the population, and if each contact with an uninfected person results in transmission of the disease. These seem like restrictions that would make it unlikely that logistic models would be good for studying epidemics, but the federal government's Centers for Disease Control and Prevention (CDC) make effective use of logistic models for projections of the yearly spread of influenza through urban populations. CDC statisticians adapt the model in a variety of ways for other types of diseases.

Logistic models are useful for tracking the spread of new technologies throughout the country. The proportion of schools in the United States that have Internet connections increased exponentially during the first half of the decade (1991–2000), then leveled off at the end, with 95 percent of the schools having Internet connections in 1999. A logistic function describes this pattern quite well. Logistic curves describe the spread of other technologies such as the proportion of families owning cell phones, the proportion of homes with computers, and the number of miles of railroad track in the country from 1850 through 1950. The logistic growth function carries a warning for companies that introduce new technologies: enjoy exponential growth in early sales, because it cannot last. When the market is saturated with the technology, new sales are very difficult to make.

Logistic models can be used to make population forecasts for anything ranging from humans to ant colonies to bacteria to fermentation levels in beer. Physicists use logistic models to study numbers of excited atoms in lasers. Agricultural chemists use logistic models to quantify the concentration of salt in soil. Bankers use the models to predict whether a person will default on a loan or credit card.

online sources for further exploration

Logistic simulations/fractals
<http://www.lboro.ac.uk/departments/ma/gallery/doubling/>
<http://mcasco.com/explorin.html>

U.S. Department of Education (National Center for Education Statistics, the Digest of Education Statistics 2000)
<http://nces.ed.gov/pubs2000/Digest99/chapter7.html>

Human population dynamics
<http://phe.rockefeller.edu/poppies/>

Logistic model of USA population
<http://www.dartmouth.edu/~math3f98/csc98/chap5/CSC.USAPop5.html>

Population
<http://www1.tpgi.com.au/users/kpduffy/logistic_t.htm>

Airport growth
<http://www1.tpgi.com.au/users/kpduffy/logistic_t.htm>

Biological growth
<http://phe.rockefeller.edu/Bi-Logistic/>

Blood pressure
<http://www.shodor.org/master/biomed/physio/cardioweb/application.html>

Electrical systems
<http://phe.rockefeller.edu/Daedalus/Elektron/>

Loglet lab
<http://phe.rockefeller.edu/LogletLab/>

National Center for Education Statistics, the Digest of Education Statistics 2000
<http://nces.ed.gov/pubs2000/Digest99/chapter7.html>

Semiconductor use
<http://phe.rockefeller.edu/LogletLab/DRAM/>

Working less and living longer
<http://phe.rockefeller.edu/work_less/index.html>

MATRICES

A *matrix* is a rectangular array of numbers. Operations that correspond to addition, multiplication, and powering of real numbers provide rules for combining matrices. Inverses of matrices correspond to reciprocals of real numbers. In addition, there are specific operations on some matrices, such as those used in game theory and graph theory that transform elements of the matrix to point the way to best decisions.

Matrices are used to solve large systems of simultaneous equations. High school students usually see matrices as a way to rewrite systems of equations. For example, $\begin{cases} 5x+3y=7 \\ 2x-y=5 \end{cases}$ can be replaced by $\begin{bmatrix} 5 & 3 \\ 2 & -1 \end{bmatrix}\begin{bmatrix} x \\ y \end{bmatrix} = \begin{bmatrix} 7 \\ 5 \end{bmatrix}$. This change seems very simple, but it generalizes the system to a matrix system $Ax = b$, which has a solution (if it exists) of $x = A^{-1}b$. The problem of accurately computing the inverse A^{-1} for large matrices is difficult, even with high-speed computer processors. This remains a critical issue for mathematicians and computer analysts, because scientists in fields as widely diverse as astronomy, weather forecasting, statistics, economics, archeology, water management, weapons races between countries, chicken production, airline travel routes, investment banking, marketing studies, and medical research rely on the efficient reduction of large matrices of information.

Matrix multiplication can provide a more secure secret code than simple replacement ciphers. Replacement ciphers (sometimes called Caesar ciphers in honor of the Roman emperor Julius Caesar, who used them in his military campaigns) encode a message by replacing each letter with another. The problem with these ciphers is that certain letters occur more frequently in languages than do others. If "z" and "m" occur most frequently in an English-language coded document, it is likely that the most frequent letter is hiding "e" and the next, "t." This one-to-one correspondence makes it easy to decode secret messages written in replacement ciphers. If the code is written with numbers that are encoded with multiplication by a matrix, the same letter encodes to different letters, depending on its position in the message. The English-language frequency distribution is then destroyed, so it is far more difficult for code breakers to decipher the message. Recipients who have the encoding matrix, however, can quickly decode the message by multiplication with the inverse of the matrix.

Some matrices describe transformations of the plane. Common geometric movements of figures, such as reflections and rotations can be written as 2×2 matrices. The table below shows some common transformation matrices.

Reflection in the y-axis	*Rotation of 90° counterclockwise*	*Reflection in the line y = x*	*Rotation of 30° counterclockwise*
$\begin{bmatrix} -1 & 0 \\ 0 & 1 \end{bmatrix}$	$\begin{bmatrix} 0 & -1 \\ 1 & 0 \end{bmatrix}$	$\begin{bmatrix} 0 & 1 \\ 1 & 0 \end{bmatrix}$	$\begin{bmatrix} \cos 30 & -\sin 30 \\ \sin 30 & \cos 30 \end{bmatrix}$

Common transformations of the coordinate plane.

Computer graphics use products of 4×4 geometric matrices to model the changes of position of moving objects in space (such as the space shuttle), transform them to eye coordinates, select the area of vision that would fit on the computer screen, and project the three-dimensional image onto the two dimensions of the video screen. The matrix products must be computed very rapidly to give the images realistic motion, so processors in high-end graphic computers embed the matrix operations in their circuits. Additional matrices compute light-and-shadow patterns that make the image look realistic. The same matrix operations used to provide entertaining graphics are built into medical instruments such as MRI machines and digital X-ray machines. Matrices such as *incidence* matrices and *path* matrices organize connections and distances between points. Airlines use these matrices on a daily basis to determine the most profitable way to assign planes to flights between different cities.

The complexity of handling the different forms of rotation that are encountered in movement requires computers that can process matrix computations very rapidly. The space shuttle, for example, is constantly being monitored by matrices that represent rotations in three-space. These matrix products control *pitch*, the rotation that causes the nose to go up or down, *yaw*, the rotation that causes the nose to rotate left or right, and *roll*, the rotation that causes the shuttle to roll over.

Stochastic matrices are formed from probabilities. They can represent complex situations such as the probabilities of changes in weather, the probabilities of rental-car movements among cities, or more simple situations, such as the probabilities of color shifts in generations of roses. When the probabilities are dependent only on the prior state, the matrix represents a *Markov chain*. High powers of the matrix will converge on a set of probabilities that define a final, steady state for the situation. In population biology, for example, Markov chains show how arbitrary proportions of genes in one generation can produce variation in the immediately following generation, but that over the long term converge to a specific and stable distribution. Biologists have used Markov chains to describe population growth, molecular genetics, pharmacology, tumor growth, and epidemics. Social scientists have used them to explain voting behavior, mobility and population of towns, changes in attitudes, deliberations of trial juries, and consumer choices. Albert Einstein used Markov theory to study the Brownian motion of molecules. Physicists have employed them in the theory of radioactive transformations. Astronomers have used Markov chains to analyze the fluctuations in the brightness of galaxies.

Ratings of football teams can be done solely on the basis of the team's statistics. But more effective and comprehensive ratings of the teams use the statistics of opponents as well. Matrices provide a way of organizing corresponding information on the team and those it has played. Solving the matrix systems that result provides a power rating that integrates information on the strength of the opponents with the information on the team. Sport statisticians contend that the use of the data make their national ratings more reliable than those that use human judgment.

Linear programming uses algorithms such as the *simplex method* to compute the most profitable solution from matrices of production. Matrices also structure inquiry into situations that have competing players with multiple choices of action. The outcomes can be organized into matrices that use players for rows and options for columns. Game theorists have developed mathematical strategies for transforming the matrices in a way that gives each player the best outcome, or in a way that avoids worst outcomes. Game strategies have been used to analyze competition for food, to determine which students get the last seats in college courses, to resolve conflicts in classrooms, and to select the best choices for potentially warring nations. The importance of game theory for the study of economic behavior is recognized by the awarding of the Nobel prize. The 1994 award in economics went to John Nash, John C. Harsanyi, and Reihard Selten for their contributions to game theory. Other Nobel awards related to game theory have been those in 1996 to William Vickrey and James Mirrlees, and to Herbert Simon in 1979. However, these were not the first Nobel prizes to recognize work with matrices. In 1973, Wassily Leontief won the prize for his prediction of best economic strategies from large input–output matrices. Leontief's theories were the basis for U.S. government policies that resulted in effective industrial production during World War II.

online sources for further exploration

Ratings of college football teams
<http://www.cae.wisc.edu/~dwilson/rsfc/rate/zenor.html>
<http://www.colleyrankings.com/#method>

David Levine's Zero Sum Game Solver
<http://levine.sscnet.ucla.edu/Games/zerosum.htm>

Cryptology and coding
<http://www.jcoffman.com/Algebra2/ch4_5.htm>
<http://www.sosmath.com/matrix/coding/coding.html>

Electronics
<http://www.math.bcit.ca/examples/ary_7_2/ary_7_2.htm>

Image rotation
<http://www.ece.gatech.edu/research/pica/simpil/applications/rotation.html>

Logging
<http://www.math.bcit.ca/examples/ary_15_2/ary_15_2.htm>

Markov chains
<http://www.sosmath.com/matrix/markov/markov.html>

Matrices in chemistry
<http://www.shodor.org/UNChem/math/matrix/>

Matrix model activities
<http://www.colorado.edu/education/DMP/activities/matrices/>

Stability of structures
<http://www.math.bcit.ca/examples/ary_5_2/ary_5_2.htm>

Transformation matrices and robotics
<http://www.math.bcit.ca/examples/ary_16_2/ary_16_2.htm>

Viewing objects in computer graphics
<http://www.math.bcit.ca/examples/ary_6_2/ary_6_2.htm>

PERIMETER

The distance around an object, or *perimeter*, is used for many purposes. The concept is used by construction workers to determine the amount of trim needed to seal the intersection between the drywall and ground, and drywall and ceiling in each room when building a house. Artists use perimeter to determine the amount of material they will need to put a frame around their pictures.

A carpenter or artist uses the concept of perimeter to build frames for pictures and paintings.

Homeowners use perimeter to determine the amount of fencing they would need for their back yard, or railroad ties to surround an outdoor patio. In an open field, a farmer can determine that the most ideal arrangement for building a rectangular pen for animals is to place his fencing in the form of a square. Suppose the farmer has 80 meters of fencing. In a rectangular pen, the unknown dimensions of the length and width can be represented by variables, l and w, respectively. The perimeter of the rectangular pen can be written as $80 = 2l + 2w$.

Rectangular pen with length l and width w.

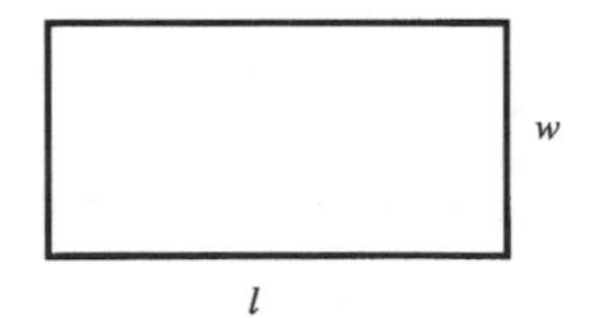

The equation can be reduced to $l + w = 40$ by dividing both sides of the equation by 2. Ideally, the farmer would like to build the largest pen so that his animals have the greatest amount of space to move around in. Thus the farmer needs to determine the dimensions that would produce a maximum area. The area, a, can be represented by the equation $a = lw$. Substituting the perimeter relationship $l = 40 - w$, the area equation can be rewritten as $a = (40 - w)w = 40w - w^2$. A graph of this function shows that the area attains a maximum value when the width is 20 meters. If the width is 20 meters, and $l = 40 - w$, then the length is also 20 meters when the area of the rectangle is a maximum value. Therefore the ideal rectangular pen based on an existing amount of fencing is a square.

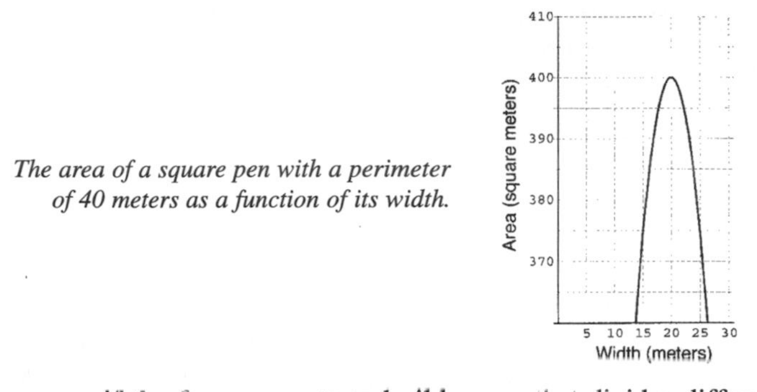

The area of a square pen with a perimeter of 40 meters as a function of its width.

However, if the farmer wants to build a pen that divides different animals, then the dimensions will have to be reconsidered. For example, suppose there are chickens and pigs in a pen that is evenly divided by a fence line. In this case, the dimensions of the most ideal pen would be determined by the equation $3w + 2l = 80$ to account for the added divider. The area of this pen is $a = (40 - 1.5w)w = 40w - 1.5w^2$. A graph of this function shows that the area attains a maximum value when the width is $40/3$ meters. If the width is $40/3$ meters, and $l = 40 - 1.5w$, then the length of the fence should be 20 meters $(40 - 1.5 \bullet \frac{40}{3})$ when the area of the rectangle is a maximum value.

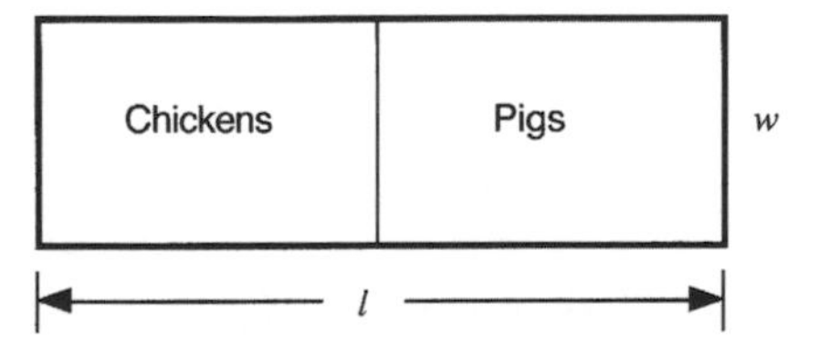

Equally divided rectangular pen with length l and width w to hold two different animals.

In addition to building fences, the concept of perimeter is used in building race tracks. For example, in track and field, a 400-meter track represents the perimeter around two congruent semicircles, where the turns and straightaways are each one hundred meters. Building a track with these dimensions requires designers and engineers to determine the distance across the track between straightaways, which represents the diameter of the semicircles. Since the circumference of a circle is π times its diameter, the circumference of a semicircle is one-half π times its diameter $(c = \frac{1}{2}\pi d)$. Rearranging the variables in the equation can show that the diameter, d, of a semicircle is $d = \frac{2c}{\pi}$, where c is the circumference of the semicircle, which is the 100-meter turn of the track. The distance across the infield of a track is $d = \frac{2 \bullet 100}{\pi}$, which is approximately 63.662 meters.

The outdoor running track at the Rock Norman Complex, Clemson University, contains two semicircles and two sides that are each 100 meters in length. The turns require precision marking to ensure that all athletes run the same distance in a race.

Marking the starting positions in different lanes uses principles of perimeter. For example, suppose the width of each lane is 1.067 meters, and the inner radius of the turn is 63.662 meters. In one lap around the track, the runner in lane 1 would run 400 meters, but the runners in the other lanes would run farther if they all started in the same position and had to stay in their lanes. The runner in lane 2 would be running around a turn with a radius of 64.729 meters, which would make each turn $\frac{64.729\pi}{2} \approx 101.676$ meters. Therefore the runner in lane 2 should start a 400-meter race ahead of the runner in lane 1 by 1.676 meters around the first turn. Since the lanes are of equal width, each runner in sequential lanes should start 1.676 meters around the first turn ahead of the previous runner. In a 4×400 meter relay, the second runner can move into the first lane after his or her first turn, which is the fifth turn overall. When staggering this relay, the runner in lane 2 should be moved 50 percent ahead of the other arrangement, since three of the turns will be run in the same lane instead of two. This means that each lane should be staggered by (1.50)(1.676) = 2.514 meters apart in this race.

online sources for further exploration

Designing a track
<http://www.crpc.rice.edu/CRPC/GT/sboone/Lessons/Titles/track.html>

Floor plans
<http://www.homebuyerpubs.com/foorplans/floorplans.htm>
<http://www.dldesigngroup.com/plans.html>
<http://ecep.louisiana.edu/ecep/math/n/n.htm>
<http://www.tnloghomes.com/homeplans/index.shtml>

Maximize the area of a rectangular field with fixed perimeter
<http://home.netvigator.com/~wingkei9/javagsp/maxarea.html>

Starting a new game farm
<http://www.agric.gov.ab.ca/livestock/elk/gamefarmapp.html>

PERIODIC FUNCTIONS

Graphs of functions that repeat shapes are called *periodic*. The horizontal length of each repetition is called the *period*. Phenomena that are based on circular motion, such as the rotation of the earth around the sun, will often result in a periodic graph. The graph below shows the hours of daylight on the fifteenth of each month for Minneapolis. The data points start with January 15th and are plotted for two years. The period for this graph is 1 year, or 12 months. The curve that has been used to approximate the data points is a sine curve, where x is the month number: Hours $= 12.2 + 2.9\sin((x - 2.3) \bullet \frac{2\pi}{12})$.

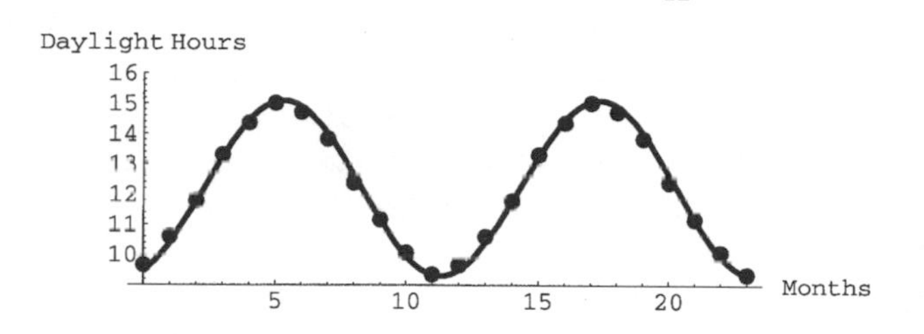

Hours of daylight for Minneapolis over a period of two years. Daylight patterns throughout a year are predictable in most cities through sinusoidal modeling.

The tilt of the earth and its rotation about the sun cause the sinusoidal pattern in hours of daylight. Because temperature in a city is dependent on hours of sunshine, plots of the average monthly temperature of American cities will be in the form of a sine curve.

Circular motion can arise from a variety of sources. The distance above ground of a passenger in a Ferris wheel produces a sine curve as the wheel rotates. Measures on a pendulum will produce periodic functions. Gravitational attraction to the moon causes tides. As the moon rotates about the earth, the heights of tides will produce a periodic function.

Sound, radar, light, radio, and ocean waves are periodic. When you press middle A on a piano, the piano strings vibrate, producing sound waves that have a period of 1/440 second. An oscilloscope provides a video screen for viewing different electrical patterns. An EKG machine in a hospital is an oscilloscope for viewing the periodic electrical patterns from a patient's heart.

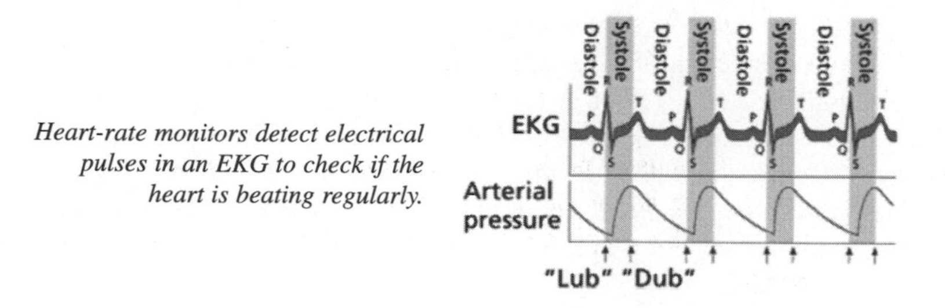

Heart-rate monitors detect electrical pulses in an EKG to check if the heart is beating regularly.

New periodic functions can be created by adding or multiplying two or more of them. You can see this when you toss two pebbles into a pond. The overlap of the waves will be a new wave. When audio speakers are arranged in an auditorium, they have to be positioned carefully so that the sound waves emerging from them do not cancel one another out or create a beat that competes with the music. Fluorescent light bulbs produce a pulsing light that sometimes adds to the cyclical refresh rate on computer screens to produce rapid light bursts that can make it hard for some people to read the computer display.

Commodity prices often follow a cyclical pattern. Hire rates for temporary-employment firms appear to form a sine curve. In biology, populations of some species of animals such as rabbits in a forest will vary over time in a cyclic pattern. When this happens, it is likely that a predator species such as a fox also has a periodic population pattern that mirrors the rabbit pattern. If there are more foxes, then there are fewer rabbits; if there are fewer foxes, then there are more rabbits. In many environments, these periodic relationships continue over time. The populations of the predator and prey do not level off.

A thermostat is an instrument used to regulate heating and cooling systems, such as an air conditioner or oven. Once a person sets the thermostat on an oven for a certain temperature, it will heat up until reaching that temperature, and then stay close to that temperature until the thermostat is changed. For example, if the oven is set to 400° Fahrenheit, it will gradually rise to that level for the first twenty minutes, and then stay at 400° until the temperature is changed or the oven is turned off. Since temperature naturally slightly varies in the air, it would also slightly change in the oven, but still oscillate near 400°. This eventual periodic function is shown in below.

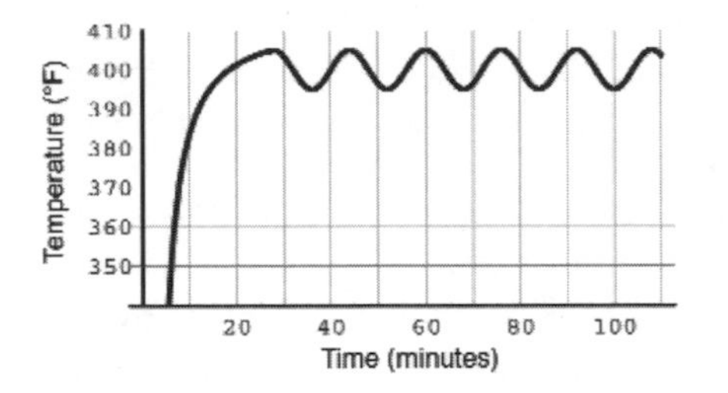

A graph of the temperature of an oven in degrees Fahrenheit as a function of time in minutes when its thermostat is set to 400°.

The cruise control in an automobile is another device that utilizes periodic behavior for a function that describes the velocity of an automobile on a highway as a function of time. As the car accelerates onto the highway, its velocity will increase and then level off near the speed that is set for the cruise control, usually the highway's speed limit. If the road has elevation changes, then the speed of the automobile in cruise control will vary slightly, since movement on hills requires different amounts of power on the automobile's engine. However, similar to a thermostat, the slight variability in speed will not affect the long-term periodic behavior of the graph describing the automobile's velocity until the brakes are touched.

The example of the electrical pulse emitted from your heart displayed on an EKG as shown in the second figure demonstrates that periodic functions do not have to be trigonometric. However, the mathematical field of Fourier transformations uses sums of trigonometric functions to approximate periodic functions of any shape.

online sources for further exploration

Art based on periodic functions
<http://www.sineart.com/>

CoolMath's links to many sites that show periodic functions
<http://www.coolmath.com/links_trig1.htm>

Alternating current
<http://www.math.bcit.ca/examples/ary_7_3/ary_7_3.htm>

Biorythms
<http://www.netcomuk.co.uk/~d_swift/biowhat.html>

Damping functions in music
<http://www.coolmath.com/dampfunction1.htm>

EKG world encyclopedia
<http://www.mmip.mcgill.ca/heart/egcyhome.html>

Heat flow
<http://www.math.montana.edu/frankw/ccp/modeling/continuous/heatflow/learn.htm>

High and low tides
<http://www.crpc.rice.edu/CRPC/GT/mwies/Lessons/lesson2.html>

Modeling with a sine function
<http://147.4.150.5/~matscw/trig/trig1.html>

Play a piano
<http://www.nws.mbay.net/maxtemp.html>

Sun or moon rise/set table for one year
<http://aa.usno.navy.mil/data/docs/RS_OneYear.html>

Temperature data
<http://www.cru.uea.ac.uk/ftpdata/tavegl.dat>
<http://www.nws.mbay.net/maxtemp.html>
<http://www.met.utah.edu/jhorel/html/wx/climate/maxtemp.html>

PLANE

A *plane* represents any flat two-dimensional surface that has infinite length and width. In everyday life, we use only finite versions of planes due to limited space and material. Walls, desktops, book covers, and floors are examples of planes. These items are made flat, because they are easier to produce and convenient to write on or cover. For example, a curved wall may be interesting to look at, but expensive to create. A flat wall, on the other hand, is much easier to wallpaper or to fix when it is damaged.

Planes have some useful properties that give people ideas about design and construction. For example, the intersection of two nonidentical planes, such as the wall and ceiling, forms a line. This idea guarantees that walls and containers made from flat surfaces can be sealed, assuming that there are no holes in them. A saw blade cuts in a straight line, because it represents two intersecting planes—the saw and the piece of wood.

Another property of planes is that three noncollinear points lie in the same plane. For example, a triangle has three vertices, so it will lie on a flat surface. Three-legged stools will never wobble, because the three ends of the legs lie on the same plane—the floor—regardless of their length. Ideally though, the leg lengths should be close to being the same to help support someone's mass near the center of the chair. Four-legged stools will sometimes wobble if one leg is longer or shorter than the other legs, because the end of one of the legs is on a different plane.

If a line or segment is perpendicular to a plane, then any congruent segments with an endpoint on that plane and another endpoint at a common point on the line or segment will be equidistant from the foot of the plane. Metal beams are placed on a radio satellite to support its receiver as waves are reflected off the dish. If they are created at the same length and intersect the receiver at the same point, then they will land on the dish at the same distance from the center. This method ensures that the beams land on the perimeter of the circle, since all points on the circle are equidistant from its center, which is directly below the location of the receiver.

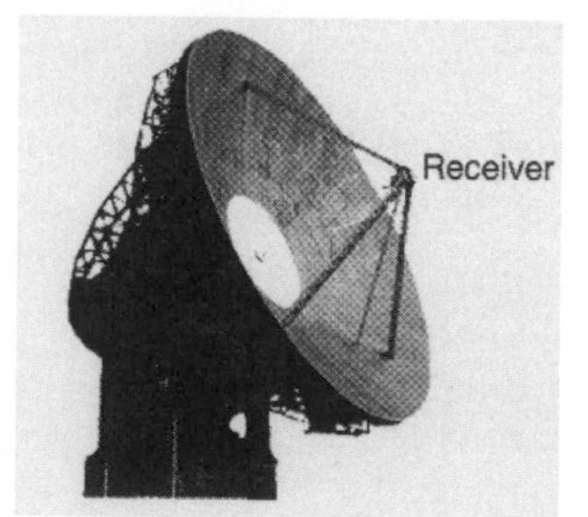

A radio antenna uses metal beams of equal length to support the receiver at its focal point.

Inclined planes—planes that are raised at an angle—are used for a variety of purposes. They are created for handicapped people in wheelchairs as an alternative to stairs. They are used to exit highways so that cars can gradually change elevation. They are also used in a variety of tools. Screws have an inclined plane that bends around its center so that they can create an angled entry when breaking a wall's surface. Screwdrivers have an inclined plane at their tip so that they can firmly fit into the top of a screw.

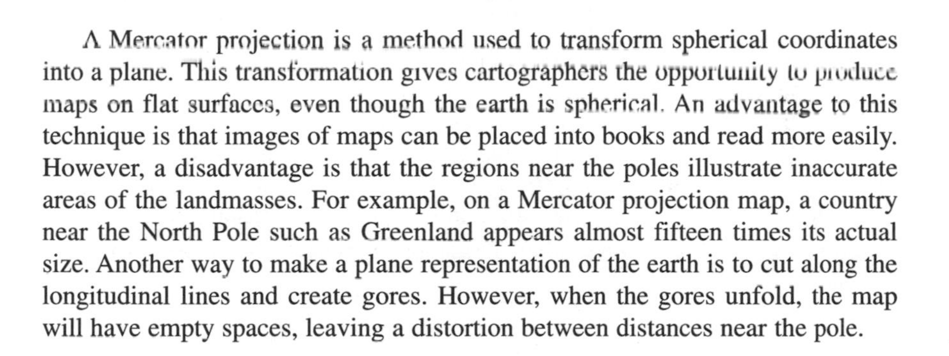

A screw uses a rotating inclined plane to drill into flat surfaces.

A Mercator projection is a method used to transform spherical coordinates into a plane. This transformation gives cartographers the opportunity to produce maps on flat surfaces, even though the earth is spherical. An advantage to this technique is that images of maps can be placed into books and read more easily. However, a disadvantage is that the regions near the poles illustrate inaccurate areas of the landmasses. For example, on a Mercator projection map, a country near the North Pole such as Greenland appears almost fifteen times its actual size. Another way to make a plane representation of the earth is to cut along the longitudinal lines and create gores. However, when the gores unfold, the map will have empty spaces, leaving a distortion between distances near the pole.

online sources for further exploration

Mercator projection
<http://www.ualberta.ca/~norris/navigation/Mercator.html>
<http://www.usgs.gov/education/learnweb/MpLesson2Act1.html>
<http://liftoff.msfc.nasa.gov/academy/rocket_sci/orbmech/mercator.html>

The three-point problem from geology
<http://jwilson.coe.uga.edu/emt725/Envir/Three.Point.html>

The wedge
<http://www.tpub.com/machines/4a.htm>
<http://www.advancement.cnet.navy.mil/products/web-pdf/tramans/bookchunks/14037_ch4.pdf>

POLAR COORDINATES

Polar coordinates locate a point by indicating a direction and a distance from a central point (the *pole*). In the point $P = [r, \theta]$, the distance r is given first, followed by the direction θ expressed as an angle of rotation from a fixed line through the pole called the *polar axis*. This is different from Cartesian or rectangular coordinates in which points are located by distances from two perpendicular axes. When the polar axis corresponds to the positive x-axis in a Cartesian plane, the Cartesian coordinates (x, y) for P can be computed as $x = r\cos\theta$ and $y = r\sin\theta$. While Cartesian graphs are rectangular, polar graphs are circular. Notice that

$$x^2 + y^2 = (r\cos\theta)^2 + (r\sin\theta)^2 = r^2(\cos\theta)^2 + r^2(\sin\theta)^2 = r^2(\cos^2\theta + \sin^2\theta) = r^2(1) = r^2,$$

which is the same as the standard equation of a circle with the center at the origin and radius r.

Polar coordinates can be used to map the earth. The figure below is a CIA (Central Intelligence Agency) map showing the northern hemisphere. The North Pole is in the center of the concentric circles of latitude. The polar axis is on the great circle of longitude that goes through Greenwich, England. On polar graph paper, this axis would be placed in the same position as the positive ray of the x-axis in Cartesian coordinates.

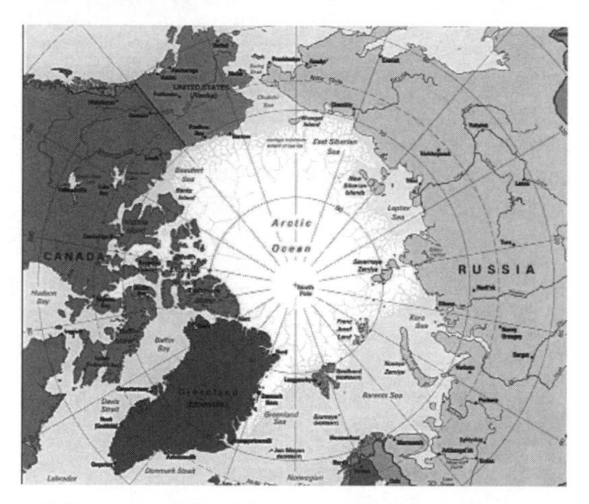

The positions and distances on the globe are represented as a polar coordinate system.

Navigators on ships and airplanes use the language of polar coordinates to specify the direction and speed of travel. Astronomers use polar coordinates to plot paths of planets and the sun with respect to a viewing position on the earth.

Polar coordinates are useful in mathematics for writing curves that cannot be written as functions or simple relations in x- and y-coordinates. The following graph of a five-leaf rose would be difficult to express in an equation using only x–y coordinates.

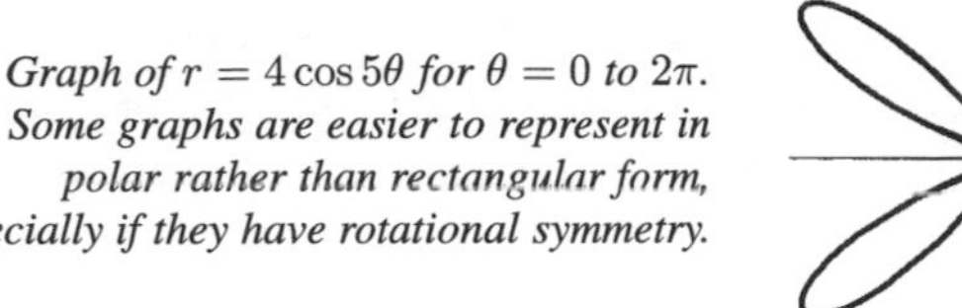

Graph of $r = 4\cos 5\theta$ for $\theta = 0$ to 2π. Some graphs are easier to represent in polar rather than rectangular form, especially if they have rotational symmetry.

Some spirals that can be graphed with polar coordinates model shapes in nature. Note in the figure below how the shape of the spiral of the form $r = ab^{\theta}$ mimics the shape of the shell of the chambered nautilus. As the creature grows, the shell compartment expands in a way that allows the nautilus to retain its shape.

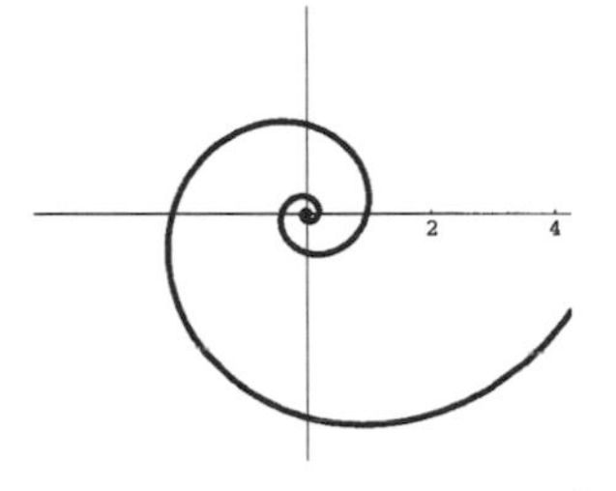

logarithmic spiral $r = 5(1.3)^{\theta}$

shell of a chambered nautilus

A nautilus resembles the polar graph $r = ab^{\theta}$.

Although polar coordinates simplify equations for some beautiful curves, they also make some equations more complicated. For example, the polar equation for the line $y = mx + b$ is $r = \frac{b}{\sin\theta - m\cos\theta}$.

Polar coordinates have surprising uses in computer graphics. The polar coordinates distortion filter available for Adobe Photoshop remaps every pixel's rectangular (Cartesian) coordinates to polar coordinates, or vice versa. This makes it easy to make objects circular as well as producing fountain-like effects associated with turning polar coordinates into Cartesian ones.

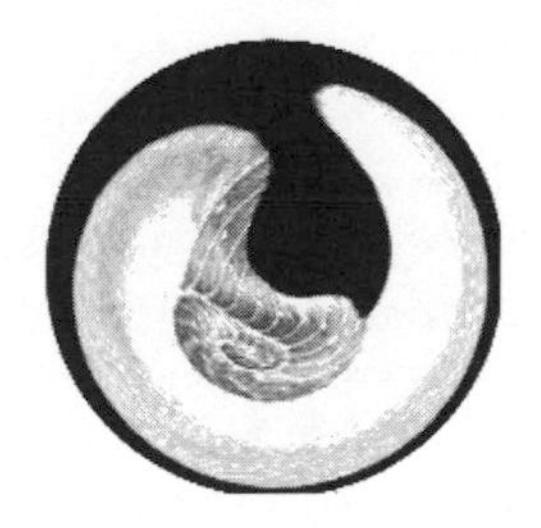

The chambered nautilus picture from the previous figure after being distorted with the polar coordinate filter in Adobe Photoshop.

Log-polar transformations have been developed to embed copyright information in computer-graphic files to preserve the copyright notice from deletion.

Spherical polar coordinates are useful in simplifying physics equations such as Schrodinger's equation and the Maxwell speed equation. In many cases, writing expressions in polar form simplifies the application of calculus and differential equation techniques.

Polar coordinates have another important application in mathematics: They simplify some operations with complex numbers. Multiplication of the complex numbers $z = a + bi$ and $w = c + di$ gives $(ac - bd) + (ad + bc)i$. The corresponding multiplication in polar form would be of the numbers $z = [r, \theta]$ and $w = [s, \phi]$. Then $zw = [rs, \theta + \phi]$. The polar form simplifies the powers and roots of complex numbers. In polar form, $z^n = [r^n, n\theta]$, which is known as *DeMoivre's theorem*. (See ***Complex Numbers*** and ***Vectors***.)

online sources for further exploration

Polar plotting and graphing
<http://mss.math.vanderbilt.edu/~pscrooke/MSS/plotpolar.html>
<http://www.world-of-newave.com/fxwavex/help/en/plug-ins/nwfxpic/polar.htm>
<http://john.redmood.com/polar.html>
<http://www.univie.ac.at/future.media/moe/galerie/zeich/zeich.html>

Spherical and cylindrical coordinates
<http://hyperphysics.phy-astr.gsu.edu/hbase/sphc.html>
<http://www.iac.tut.fi/~sahrakor/research/teksti/node8.html>
<http://www-istp.gsfc.nasa.gov/stargaze/Scelcoor.htm>

Logarithmic spirals
<http://brand.www.media.mit.edu/people/brand/logspiral.html>
<http://www.notam.uio.no/~oyvindha/loga.html>
<http://www.meru.org/goldmean.html>
<http://xahlee.org/SpecialPlaneCurves_dir/EquiangularSpiral_dir/equiangular Spiral.html>

Sun position in polar coordinates
<http://www.jgiesen.de/sunpol/index.html>

Azimuth and elevation
<http://www-istp.gsfc.nasa.gov/stargaze/Scelcoor.htm>

Computer graphics
<http://www.adscape.com/eyedesign/photoshop/four/filters/polarcoordinates.html>
<http://www.asahi-net.or.jp/~nj2t-hg/ilpov21e.htm>
<http://www.blueberry-brain.org/syndyn/spirals/figsfrac.htm>

Polar coordinates in robotics
<http://www.math.bcit.ca/examples/ary_16_1/ary_16_1.htm>

Polar distortion filter
<http://www.adscape.com/eyedesign/photoshop/four/filters/polarcoordinates.html>

POLYNOMIAL FUNCTIONS

A *polynomial function* $f(x)$ has a general equation $f(x) = a_1x^n + a_2x^{n-1} + a_3x^{n-2} + \ldots + a_z$, where coefficients and constants are associated with a and exponents associated with n are positive integers. Linear functions, such as $y = 3x - 5$, and quadratic functions, such as $r = 3w^2 - 5w + 7$, are polynomial functions that have numerous applications discussed elsewhere in this book (see ***Linear Functions*** and ***Quadratic Functions***). Polynomial functions with degree three or greater are found in applications associated with volume and financial planning.

Empty open-faced containers such as crates are put together by attaching a net of five rectangles. A rectangular piece of plastic can be cut so that it can turn into an open-faced rectangular prism when folded at its seams. If a square piece is cut out of each corner of a rectangle, then four folds will form a net with five rectangles that can be formed to develop the prism, as shown below.

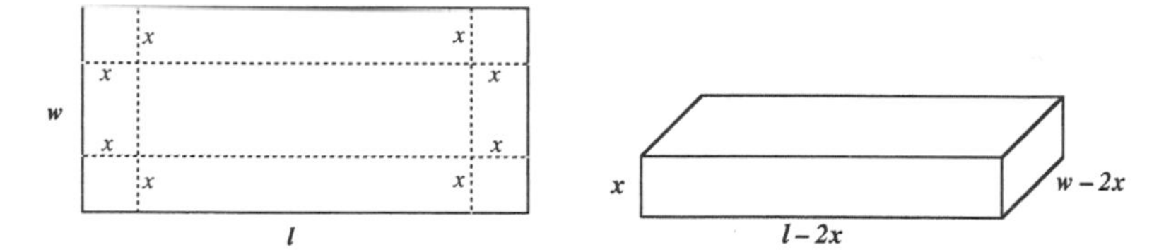

Open-faced prism with dimensions x by $l - 2x$ by $w - 2x$ formed by cutting squares with side length x out of the corners of rectangular sheet with dimensions l by w.

A manufacturer is probably interested in finding the location to cut the square from the corners so that the consumers will be able to fill the crate with the most amount of material. In essence, the goal is to maximize the volume based on a fixed amount of material. Suppose that square corners are removed from a rectangular sheet of plastic with dimensions of 6 feet by 4 feet. Each side of the prism can be represented in terms of the edge length, x, of the square that was removed from the corners, as shown below.

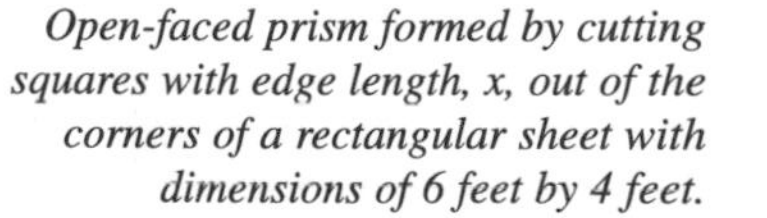

Open-faced prism formed by cutting squares with edge length, x, out of the corners of a rectangular sheet with dimensions of 6 feet by 4 feet.

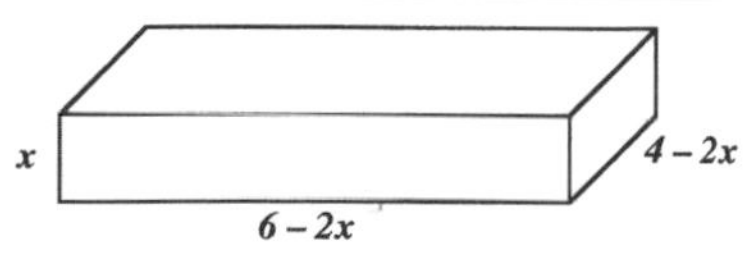

The volume of the crate, v, is the product of its dimensions, so it can be represented by the equation $v = x(6 - 2x)(4 - 2x)$. This equation is a polynomial function, because it is the factored form of $v = 4x^3 - 20x^2 + 24x$. A relative maximum of this function on a graph, as shown on the following page, within a domain between 0 and 2 feet occurs when $x \approx 0.78$ feet, or about 9.4 inches. This means that the crate with the largest possible volume will occur when squares with an edge length of 9.4 inches are cut from the corners.

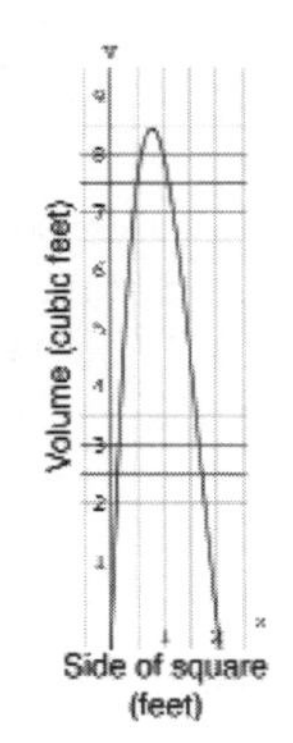

Graph of the volume of a prism formed by folding a sheet of paper with squares of edge length, x, removed from each of the corners.

Long-term investing uses a polynomial function to account for money that is invested each year. Suppose an account was set up so that you contributed money each year towards your retirement based on a fixed percentage of interest, assuming that you continued to add a minimum amount of money to the account each year and did not withdraw money at any time. The total amount of money, m, in the bank after n years based on an annual interest rate of r percent can be represented by the function

$$m = a_1(1 + \tfrac{r}{100})^n + a_2(1 + \tfrac{r}{100})^{n-1} + a_3(1 + \tfrac{r}{100})^{n-2} + \ldots + a_z,$$

where the coefficients, a, are the individual amounts of money deposited into the account after each year. For example, if \$500 is deposited at the end of the first year, \$700 at the end of the second year, \$800 at the end of the third year, and \$400 at the end of the fourth year, then the total amount of money in the account at the end of the fourth year is determined by the equation

$$m = 500(1 + \tfrac{r}{100})^3 + 700(1 + \tfrac{r}{100})^2 + 800(1 + \tfrac{r}{100}) + 400.$$

This means that the initial deposit of \$500 will compound three times, the second deposit of \$700 will compound two times, and so on. If an employee uses this retirement plan for only four years and wants to know the value of the account 21 years after the first investment, then the equation would be rewritten to

$$m = 500(1 + \tfrac{r}{100})^{20} + 700(1 + \tfrac{r}{100})^{19} + 800(1 + \tfrac{r}{100})^{19} + 400(1 + \tfrac{r}{100})^{18}.$$

This information is useful for people in their financial planning so that they can learn how to save money for their children's education and their own retirement.

online sources for further exploration

Antenna pattern correction
<http://earth.esa.int/0xc1cce41c_0x00005bfe>

Application of polynomial functions
<ftp://cq-pan.cqu.edu.au/pub/smad/senior/mathsb/mb_if005.doc>

Building boxes
<http://www.pbs.org/teachersource/mathline/lessonplans/hsmp/buildingboxes/buildingboxes_procedure.shtm>

Drag racing
<http://ceee.rice.edu/Books/CS/chapter3/data1.html>

Shrimp
<http://144.35.21.240/mathdept/labs/shrimp.htm>

Toxic waste
<http://www.eddept.wa.edu.au/centoff/graphcalc/tasks/ic6pc.pdf>

▲▼▲

PROBABILITY

A *probability* is a number between 0 and 1 that tells us how likely an event is to happen. Probabilities are expressed as ratios or percents. When probabilities are computed from an analysis of possible outcomes, such as the probability that a sum of 7 will show on the toss of a pair of dice, the probability is sometimes called a *theoretical probability*. When the probability is computed on the basis of experience or surveys, such as the probability of a randomly selected adult being a smoker, it is called a *relative frequency* or *experimental probability*. Gambling probabilities are sometimes expressed as *odds*. Odds express the ratio of successes to failures, or vice versa. If you hear a bettor on a horse race say, "The odds against Fleetfoot winning are five to two," that means for every five times Fleetfoot loses, he will win twice. The probability of losing is five out of seven, and the probability of winning is two out of seven.

Games that have equal probabilities for all participants are called *fair*. Most people think that tossing a coin to determine who kicks off a football game is fair, because their life experience with coins indicates that the probability of a head is equal to a probability of a tail. Determining the winning state-lottery ticket by pulling winning ping-pong balls from agitated buckets is felt to be fair, because every number has an equal chance of being selected. This perception has been written into the election laws of many states. Illinois determines the seventh member of its redistricting committee (otherwise evenly divided between Democrats and Republicans) by pulling a name out of a hat. In 1998, the tie in ballots for mayor of Duluth, Minnesota, was broken by the toss of a coin. Kansas once settled a tie for state representative by having the two candidates pull chips from a bag that contained six black and six white backgammon chips. The winner was the first to draw a white chip. New Mexico allows tied candidates to play any game of chance to break a tie. Coin tossing remains the favorite way of breaking ties in the state, but as recently as 1998, two candidates who were tied in the election for mayor of Estancia, New Mexico, opted to play five-card stud for the job.

Statisticians use probabilities and relative frequencies in determining relationships. A key concept from probability is the idea of *independence*. The formal mathematical definition of independence is given in the equation $P(A + B) = P(A) \bullet P(B)$. If two events, A and B, are independent, then the probability of them both happening is the product of the separate probabilities. The concept of independence is behind investigations such as the Physicians' Health Study, which tested the effects of aspirin on over 22,000 doctors. Half the doctors were given a daily dose of aspirin, and half were given a neutral pill (placebo). Doctors didn't know what kind of pill they received. The researchers periodically contacted the participants to find out if the physician had suffered a heart attack. The results showed that 0.9 percent of the participants who received aspirin had heart attacks, and 1.7 percent of those with the placebo had heart attacks. Although it looks like the percents favor aspirin, the percents are so small that it is possible they were due to chance. The study data is shown on the following table. This is called a *contingency table*.

	heart attack	***no heart attack***	TOTAL
aspirin	104	10,933	11,037
placebo	189	10,845	11,034
TOTAL	**293**	**21,778**	**22,071**

The results from the aspirin–physician heart study.

Statisticians assume that the medication and heart attacks are independent. If so, then P(aspirin and heart attack) = P(aspirin) • P(heart attack). Using the relative frequencies from the total row and total column gives the following product: $\frac{11037}{22071} \bullet \frac{293}{22071}$. That probability times the number of participants tells how many doctors receiving aspirin would have had heart attacks if heart condition were independent of medication. That frequency is 147. As you can see from the table below, the almost equal separation of physicians into aspirin and placebo treatments indicates that the expected values for heart attacks should have been separated into almost equal proportions.

	heart attack	***no heart attack***	TOTAL
aspirin	147	10,890	11,037
placebo	146	10,888	11,034
TOTAL	**293**	**21,778**	**22,071**

Expected frequencies for the aspirin–physician heart study The computations assume that the totals represent the population, and that heart condition is independent of medication.

The statistician conducts a *chi-square test* to compare the actual frequencies to the expected frequencies. In this case, the chi-square indicated that the observed frequencies were not close to the expected values, so aspirin reduced heart attacks.

Making careful lists and working from simple examples can determine many probability problems. How many families with three children have exactly two boys? If boys and girls are equally likely, you can list eight possibilities: *BBB*, *BBG*, *BGB*, *BGG*, *GBB*, *GBG*, *GGB*, and *GGG*. The list is called the *sample space*, because each family is equally likely. Three of these, *BBG*, *BGB*, and *GBB*, represent two boys and one girl. So the probability of a family of three children having exactly two boys is three-eights, or 37.5 percent.

The problem of finding how many families would have two boys in three children can be approached through a *simulation.* A simulation replaces the elements of this problem with repeated trials of an experiment using objects that behave like the birth of children. Tossing a coin could represent the birth of a child. If you were to determine boys by the head of the coin showing, you could simulate a family of three children by tossing three coins, say a penny for the first child, a dime for the second child, and a quarter for the third. This experiment can be carried 500 or more times very quickly. The probability of two boys would be estimated by the proportion of times the three coins showed exactly two heads. In one experiment, this proportion turned out to be 35.8 percent, which is a little less than the value computed from the sample space. It is now common to use computers to model complex relationships with simulations. Computers can generate random numbers (or numbers that act randomly) and perform rapid computation of probabilities. The Defense Department uses simulations to evaluate outcomes of military actions. Aircraft designers use computer simulations of air molecules hitting the surface of an airplane to determine its most efficient shape. The Centers for Disease Control uses simulations to predict the paths of epidemics. It makes recommendations for vaccinations and prevention procedures based on the outcomes of its simulations.

Coins and children present examples of *binomial probability* situations. When there are two outcomes of a single trial (heads or tails on one coin, boy or girl for one birth), and a fixed number of independent trials, the computation of outcome probabilities can be generated by terms in the expansion of the binomial $(p+q)^n$, where n is the number of trials, p is the probability of one outcome (called the success), and $q = 1 - p$ is the probability of failure. Families of three children would be modeled by

$$(p+q)^3 = p^3 + 3p^2q + 3pq^2 + q^3.$$

The term $3p^2q$ would represent the probability of two boys and one girl. Since $p = q = \frac{1}{2}$, the value $3p^2q = \frac{3}{8}$ agrees with our previous computation.

The binomial probability theorem provides direct solutions for problems that don't have equal probabilities such as the proportion of recessive genes in a population or how many people should be booked for flights so that there are no empty seats. In a situation in which there are different percentages of a dominant gene *A* and a recessive gene *a*, shouldn't the dominant gene eventually "win out" in the population? For example, 1 out of 1,700 Caucasian children is born with cystic fibrosis, which is caused by a recessive gene. Unfortunately, that proportion remains the same from generation to generation. In 1908, a British mathematician and a biologist used binomial probabilities to explain genetic stability. The *Hardy-Weinberg equation* models the genetic distribution with the perfect square binomial $(p+q)^2 = p^2 + 2pq + q^2$, where p is the proportion of the dominant gene A, and q is the proportion of the recessive gene a. In the situation of cystic fibrosis, p^2 is the proportion of people who are pure dominant, $2pq$ is the proportion of people who do not have cystic fibrosis but are carriers, and q^2 is the proportion of people who have cystic fibrosis. In Caucasian children, q^2 is the incidence rate of 1/1700 or 0.00059. Taking the square root gives $q = 0.024$. The recessive gene a for cystic fibrosis accounts for $q = 2.4$ percent of the genes, and the dominant gene A accounts for $p = 97.6$ percent. Computing the proportion of people free from the cystic fibrosis gene gives $p^2 \approx 0.9253$, and the proportion of people who are carriers of the recessive gene is $2pq \approx 0.0468$. About 92.5 percent of the population is free of the cystic fibrosis recessive gene, but 4.7 percent are carriers. In the absence of mutations and migration, these proportions will remain constant from generation to generation. *Markov chains* can be used to handle the relative frequencies of many species in populations as well as gene pairs. (See ***Matrices***.)

Airline scheduling can be considered a binomial probability problem. Assume that 90 percent of the people who buy tickets actually show up at the airport to board the plane. If the plane seats 50, then on average, 90 percent of 50 seats = 45 would be filled. Airlines run on small profit margins, so those five empty seats could make the flight a money loser. Airlines attempt to solve this problem by selling more than 50 tickets for the flight. If they sold 52 seats, for example, on average, 47 people would actually show up for the flight. But there would be times when 51 or 52 people showed up. Some people would not get on the flight, so the airline would have to pay a penalty and incur the wrath of the passengers who had a ticket but did not get a seat. The airlines want to oversell just enough to regularly fill all seats, but not to overbook so much that the penalties outweigh the additional ticket income. The binomial expansion $(p+q)^{52}$ will give the chances that one or more ticketed customers will lack a seat. The expansion of $(p+q)^{52}$ starts out as $p^{52} + 52p^{51}q + 1326p^{50}q^2 + \ldots$. The first term gives the probability that all 52 ticketed passengers will show up, $(0.9)^{52} \approx 0.00417$. The second term gives the probability that 51 ticketed passengers will show up, $52(0.9)^{51}(0.1) \approx 0.000463$. Adding these probabilities gives 0.0046. About five flights in every thousand will have customers who would not get seats. This isn't a big probability, so the airline would be safe in selling 52 seats for flights on this size plane. With these probabilities, the airline could compute the expected profits on its flights, accounting for the penalties paid to the unserved passengers. (See ***Expected Value***.) They would have to repeat the computation for 53 tickets sold, 54 tickets, and so on. At 55 tickets sold, for example, the binomial expansion indicates that about one-third of the flights would have to turn away ticketed passengers. That is probably too often. Larger powers of binomial situations can be estimated with normal distributions. (See ***Standard Deviation***.)

Probability arguments are common in court cases. DNA (deoxyribonucleic acid) matching gives a probability that blood, semen, or hair found at a crime scene matches the accused. The early cases of prosecution based on DNA matching produced lengthy arguments about the accuracy of the techniques and the computation of the probabilities. For example, some of the genetic markers that are used in establishing probabilities occur in different proportions in different racial groups. By 1996, recommendations from the National Research Council, the National Institute of Justice, and other government and legal organizations resulted in standardized laboratory techniques and computations of probabilities, so DNA evidence is as well accepted as fingerprint matches.

Probabilities have been used to determine whether juries were representative in gender and racial composition to the communities they served. Lawyers for Al Gore and George W. Bush used probability arguments before Florida courts to persuade judges that their respective parties should prevail in the contested presidential election of 2000. In the Microsoft antitrust case, the Department of Justice presented probabilities that the Microsoft Corporation would force other companies out of business. Courts have based financial awards to patients whose cancer was misdiagnosed by doctors on computations of the patients' reduced probability of survival.

Probabilities can be computed from geometry formulas. Consider balls falling uniformly on a square piece of cardboard 20 inches on a side that has a circular hole 5 inches in diameter. The proportion of balls that fall through the hole is proportional to the ratio of area of hole to area of the cardboard. This would be computed using the formulas for area of circle and square: $\frac{\pi(2.5)^2}{20^2} \approx 0.049$. A ball has about a 5 percent chance of falling through the hole rather than bouncing off the cardboard. Winning carnival games is much more difficult than it appears!

online sources for further exploration

The geometry junkyard shows geometric probability problems
<http://www.ics.uci.edu/~eppstein/junkyard/random.html>

The birthday problem
<http://www.mste.uiuc.edu/reese/birthday/>

Discrete probability
<http://www.colorado.edu/education/DMP/activities/discrete_prob/>

Diffusion
<http://www.math.montana.edu/frankw/ccp/modeling/probability/diffusion/learn.htm>

Lottery odds calculations
<http://www.lottery.state.mn.us/odds.html>
<http://www.howstuffworks.com/lottery1.htm>
<http://www.alllotto.com/oddscalc.asp>
<http://indigo.ie/~gerryq/Lotodds/lotodds.htm>

Nuclear medicine
<http://www.math.bcit.ca/examples/ary_11_8/ary_11_8.htm>

Poker probabilities
<http://www.pvv.ntnu.no/~nsaa/poker.html>

Probability and utility of the real world
<http://research.microsoft.com/~horvitz/real.htm>

Probability in the real world
<http://forum.swarthmore.edu/dr.math/faq/faq.prob.world.html>

▲▼▲

PROPORTIONS

Proportions are equations that compare ratios or scaled quantities. Cartographers use proportions to make maps, because they need to scaledown distances so that large pieces of land can be viewed on a sheet of paper. For example, the state of Illinois is approximately 370 miles long. If the map maker wants to place Illinois on a sheet of paper that is 25 cm long, a proportion that can be used to determine a scale in this situation is $\frac{s \text{ miles on map}}{370 \text{ miles}} = \frac{1 \text{ cm on map}}{25 \text{ centimeters}}$. Cross multiplying these quantities helps solve the equation, $s = 370/25 = 14.8$. The legend on the map might indicate that 1 cm represents 15 miles.

Eratosthenes used a proportion to determine the radius of the earth in around 230 BC. He had traveled between the cities of Alexandria and Cyrene and determined that its distance was around 5,000 stades, where each stade is about 559 feet. At noon, he had measured the angles of shadows formed by sticks in the ground and determined that there was not any shadow at Cyrene, and an angle of elevation of $\alpha = 82.8$ degrees at Alexandria, as shown below.

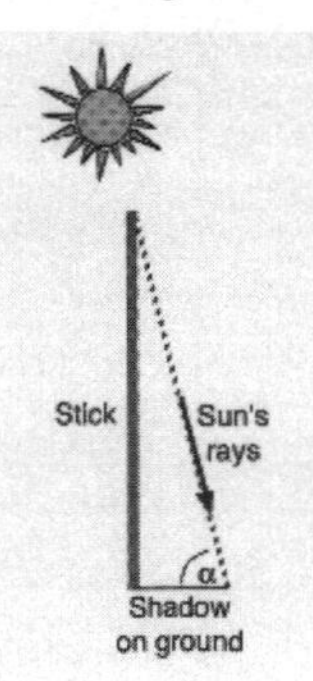

The shadow produced on the ground at noon at Alexandria based on an angle of elevation of α degrees.

Eratosthenes argued that the angle formed near the top of the stick, 7.2 degrees, is the same as the central angle in the earth that determines the sector between the two cities, since light rays travel parallel to the earth, as shown as follows.

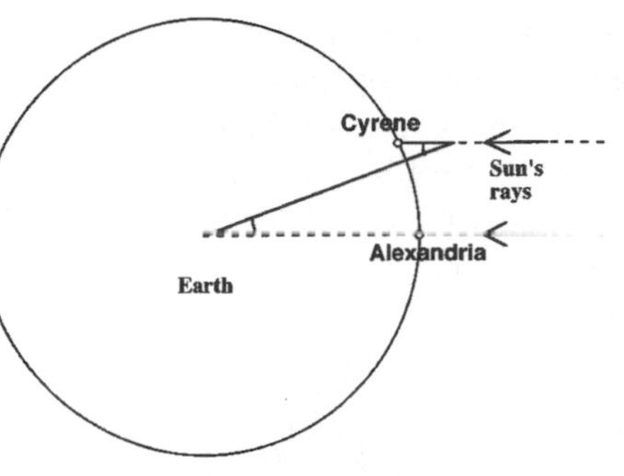

Congruent angles formed by the sun's rays at Cyrene and at the center of earth. Note that figure is not drawn to scale.

Therefore the distance from Cyrene to Alexandria represented 7.2/360 of the earth's circumference, since there are 360 degrees in a circle. Using the proportion,

$$\frac{\text{7.2 degrees between cities}}{\text{360 degrees in circle}} = \frac{\text{5000 stades between cities}}{\text{number of stades around earth}},$$

Eratosthenes determined that the distance around the earth is about 250,000 stades, or 139,750,000 feet. Since the cross section of the earth is approximately a circle, the radius can be determined by using the equation $c = 2\pi r$, comparing the circumference, c, to its radius, r. Substituting the value c = 139,750,000 feet and solving the equation will show that the radius of the earth is about 22,241,900 feet, or 4,214 miles. That's only 6 percent off the actual distance of 3,963 miles!

Movie screens are designed to be in a similar proportion as the dimensions of film cells.

Movie screens are created to handle film with specific dimensions. Suppose the width of each film cell is 5.48 cm and the height is 2.30 cm. An ideal movie screen would show the entire picture without cropping out any of the sides. A small movie theater may leave a horizontal length of 7 meters, about 23 feet, to place its screen. In order to project the film perfectly on the screen, a proportion comparing the height and width needs to be used so that the correct height of the screen can be accurately determined. The height of the screen, h, is equal to approximately 2.94 meters, or 9.66 feet, by solving the equation determined by the proportion

$$\frac{h \text{ vertical meters on screen}}{\text{2.30 vertical cm on film}} = \frac{\text{7 horizontal meters on screen}}{\text{5.48 vertical cm on film}}.$$

Proportions are used to predict the mass of a dinosaur with scaled models. Since a model is a miniature version of the actual dinosaur, paleontologists use the ratio

$$\frac{\text{length of actual dinosaur}^3}{\text{length of model}^3} = \frac{\text{volume of actual dinosaur}}{\text{volume of model}}.$$

This ratio is cubed, because volume is a three-dimensional concept, compared to a one-dimensional concept of length. For example, the volume of a cube is the length of one of its edges raised to the third power. The ratio is used to find the volume of the actual dinosaur, since the other measurements can be taken from a model and a fossil of the skeleton. (See ***Similarity*** for an example.) The density of an object is the ratio of its mass to volume, so the mass of the dinosaur can be predicted by estimating the density of a dinosaur as that of a modern-day reptile or mammal.

A scaled model of a tyrannosaurus can be used to predict its actual mass with information about its actual length.

The population of wildlife animals is determined by tagging animals and using proportions. It is important to know these populations in order to understand if a species is at risk of endangerment, or if there is an overpopulation that is affecting an ecosystem. Every animal in a region cannot be counted, because it would be too difficult to find all of them, not to mention that it would be distracting and potentially disturbing to the ecosystem if ecologists were constantly roaming around. Consequently, a scientist will go to an area such as a forest and temporarily capture animals to tag them. In addition to placing tags on them, the scientist will likely examine their health to understand their potential to reproduce or spread disease. A few weeks later, after the animals have had a chance to roam around the forest, the scientists will recapture a group of animals again to check their health and keep track of the proportion of animals that are tagged. This information will help the scientists determine how many animals are in that region of the forest, assuming that the birthrate and death rate are fairly similar during that time period. This method of estimating animal populations is called *capture–recapture*.

It is difficult to track animals due to their mobility and ability to hide well. A capture–recapture method can be used to tag and predict their population without having to count them all.

For example, suppose 50 deer are captured and tagged in a forest. Two weeks later, 100 deer are captured, and 18 of them have tags. The proportion

$$\frac{\text{number of tags in population}}{\text{number of tags in second sample}} = \frac{\text{number of deer in population}}{\text{number of deer in second sample}}$$

can be used to predict the total number of deer in that region of the forest. In this case, the number of deer in the population, p, is approximately 277 based on a solution to the equation $\frac{50}{18} = \frac{p}{100}$.

In baseball, an earned run average (ERA) is a statistic that describes how many runs a pitcher would be expected to give up in a nine-inning game. However, a pitcher hardly lasts that long during a single game. On a good day, a major league pitcher will play for about six to nine innings. However, if the pitcher is giving up a lot of runs he will play considerably less, perhaps from one to five innings. Regardless of the length of the pitching performance, an ERA is used to compare the different pitchers. Typically, better pitchers have lower ERAs. For example, a pitcher that gives up two runs in seven-and-a-third innings has an ERA of 2.45, because the ERA is determined by the proportion

$$\frac{\text{number or runs allowed}}{\text{number of innings pitched}} = \frac{\text{number of runs in an entire game (or ERA)}}{\text{number of innings in a nonextended game}}.$$

Substituting the numbers, the proportion is $\frac{2}{7\frac{1}{3}} = \frac{\text{ERA}}{9}$. A bad pitcher may give up five runs after two-and-two-thirds innings and have an ERA of 16.88. Usually, the ERA is a statistic that represents a player's performance over an entire season, and is updated after each pitching performance.

In 1619, Johannes Kepler determined a proportion relating the mean distances, d, that planets were from the sun and the period of their revolution, p. This proportion, $\frac{d_1^3}{d_2^3} = \frac{p_1^2}{p_2^2}$, was determined through data collection, and can be proven using Newton's theory of gravitation. At the time, this information was helpful to astronomers to predict the approximate distance that another planet is from the sun. For example, Mars is observed in the night sky to have a period of 687 days revolving around the sun. There are 365 days in an earth year, so the ratio of periods of these orbits is 1.882. Kepler's proportional formula can be rewritten as $\left(\frac{d_1}{d_2}\right)^3 = \left(\frac{p_1}{p_2}\right)^2$ because exponents distribute in any expression involving a quotient of two numbers. When the Mars-to-earth ratio is substituted into the equation, the ratio of the distances will be approximately 1.524, as a result of the solution to the equation $\left(\frac{d_1}{d_2}\right)^3 = (1.882)^2$. This means that Mars is about 50 percent further from the sun than the earth, perhaps one reason in understanding why most of Mars regularly maintains temperatures below 0° Fahrenheit. (See ***Inverse Square Function*** for another explanation.) This information is also helpful for astronomers to predict when a spacecraft can be launched from earth so that its trajectory would come in close contact with a planet. For example, the *Voyager* ships launched in 1977 had trajectories that placed them near Jupiter, Saturn, Uranus, and Neptune in order to take photos that could be sent back to earth for further study.

The strength of an animal or object is proportional to its surface–area-to-weight ratio. Small insects can carry objects much greater than their mass, while humans can only carry small percentages of their mass. If an ant were to grow in size, it could not maintain its surface area-to-weight ratio. Suppose a giant ant were twenty times longer than a tiny ant. Since area is related to the square of length in an object, the ant's surface area would increase by a factor of 20^2, or 400. The volume of the ant, which is proportional to its mass, would increase by a factor of 20^3, or 8,000, because volume is related to the cube of an object's length. If the original ratio $\frac{\text{surface area}}{\text{volume}} = \frac{s}{v}$ for the ant, the giant version 20 times bigger in each dimension would have $\frac{\text{surface area}}{\text{volume}} = \frac{400\ s}{8000\ v}$. Hence the relative strength of a giant ant would be only $\frac{400}{8000}$ or 5 percent as much as a tiny ant. For example, if a regular ant can carry ten times its body weight, then a giant ant could only carry one-half its body weight. If such a giant ant existed, it would probably have slightly different proportions than the smaller ant, since the cross-sectional area of its legs would probably need to be proportionally larger in order to maintain a dramatic increase in mass. Otherwise, there would be close to 90 times ($20^{3/2}$) as much pressure on its legs than before, which would probably cause the legs to snap. Ouch! That is why elephants need tree-trunk-style legs in order to support their own weight. This proportional understanding of strength helps designers build stronger paper towels, bags, and boxes, and helps engineers build stronger and more durable machines that can withstand pressure such as airplanes and bridges.

online references for further exploration

Eratosthenes of Cyrene
<http://share2.esd105.wednet.edu/jmcald/Aristarchus/eratosthenes.html>

Build a solar system
<http://www.exploratorium.edu/ronh/solar_system/>

Circumference of earth using techniques by Eratosthenes
<http://share2.esd105.wednet.edu/jmcald/Aristarchus/eratosthenes.html>
<http://w3.ed.uiuc.edu/noon-project/>

Earned Run Average all-time leaders
<http://www.baseball-almanac.com/piera1.shtml>
<http://www.baseball-almanac.com/piera4.shtml>

Nuclear medicine
<http://www.math.bcit.ca/examples/ary_11_1/ary_11_1.htm>

Orbit simulation
<http://observe.ivv.nasa.gov/nasa/education/reference/orbits/orbit3.html>

Map making
<http://www.sonoma.edu/GIC/Geographica/MapInterp/Scale.html>
<http://www.epa.gov/ceisweb1/ceishome/atlas/learngeog/mapping.htm>

Proportional representation in voting
<http://www.ci.cambridge.ma.us/~Election/pr-quota.html>
<http://www.ci.cambridge.ma.us/~Election/ballots.html>

Scale models
<http://www.faa.gov/education/resource/f16draw.htm>
<http://www.pbs.org/wgbh/nova/pyramid/geometry/model.html>
<http://www.americanmodels.com/sscale.html>

Understanding scale speed in model airplanes
<http://www.astroflight.com/scalespeed.html>

▲▼▲

PYTHAGOREAN THEOREM

The *Pythagorean theorem* states that the sum of the squares of the legs of a right triangle, $a^2 + b^2$, is the same as the square of its hypotenuse, c^2. There are over 100 proofs of the Pythagorean theorem, many of which show that the sum of the areas of squares on the legs is equal to the area of the square on the hypotenuse, as shown in the figure below. Conversely, any triangle that has sides that are related by the equation $a^2 + b^2 = c^2$ must have a right angle opposite the longest side.

The Pythagorean theorem illustrates that the sum of the areas of the squares connected to the legs of a right triangle is equal to the area of the square connected to the hypotenuse of a right triangle.

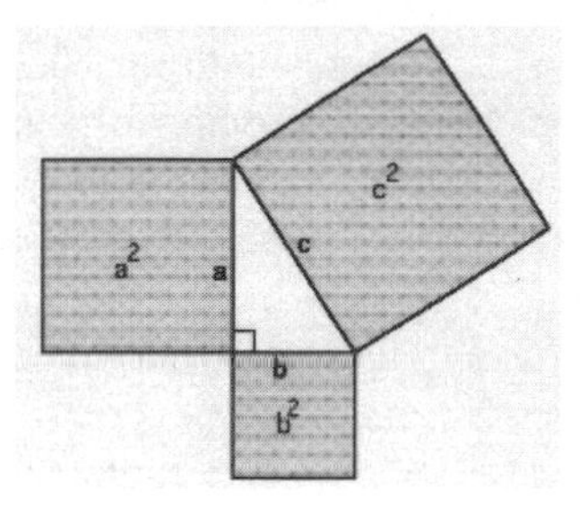

The Pythagorean theorem is useful on a baseball diamond for several reasons. Since the bases are each 90 feet apart in the form of a square, the theorem helps us find the distance the catcher has to throw the ball to second base when a runner is trying to steal. The right triangle formed would be with half of the infield, where the legs of the triangle are the base paths of 90 feet each, and the hypotenuse is from home plate to second base. The hypotenuse can be found by solving the equation $90^2 + 90^2 = c^2$. Solving for c will show that the throw is about 127.3 feet. This information is useful, because it will give coaches an idea about how hard the catcher needs to be able to throw a ball accurately in order to throw a runner out. If the catcher throws a ball at about 70 miles per hour, then it will only take about one-and-a-quarter seconds for the ball to reach the base.

The geometry of rhombuses and the Pythagorean theorem can be used to show that the center of the pitcher's mound is not in the pathway of the ball when it is thrown from third to first base. The diagonals of the square running-path between the bases are perpendicular bisectors of each other, forming congruent right triangles in the center. If the pitcher was placed at the intersection of the diagonals, he might get hit by a throw from the third baseman. To avoid contact, the pitcher needs to be placed closer to home plate than this intersection. The Pythagorean theorem gave the distance between home and second base to be 127.3 feet. The pitcher must be closer to home plate than 63.6 feet. The actual placement of the center of the pitching mound is 60.5 feet from home plate.

The Pythagorean theorem is used to approximate the distance of two nearby towns on a map. Changes in the earth's curvature are minimal within short ranges, so the latitude and longitude positions can serve as points on a coordinate plane. For example, suppose that Smithsville is five miles north and two miles east of Laxtown. The two cities would be 5.39 miles away on a map, representing the distance that the "crow flies." This distance can be determined by solving the equation $5^2 + 2^2 = d^2$ that is determined with the Pythagorean theorem.

Carpenters use Pythagorean triples to verify that they have right angles in their work. For example, a carpenter making a cabinet can perfectly align pieces of wood in a right angle with the use of only a tape measure. Using the Pythagorean triple {3,4,5}, or any multiple such as {12,16,20}, the carpenter can place a mark on the bottom after 12 inches, a mark on the side after 16 inches, and rotate the intersecting boards at its hinge until the distance between the markings is 20 inches. A triangle with sides of 12, 16, and 20 inches is a right triangle, since $12^2 + 16^2 = 20^2$.

Construction workers building along the sides of mountains use the Pythagorean theorem to determine the amount of supplies needed to create a railroad track for a funicular or a cable line for a gondola. The horizontal and vertical distances from the foot of the mountain to its top can be determined on a map, forming the legs of a right triangle that can be drawn in the mountain's center. The third side of the triangle, the hypotenuse, represents the walk up the mountain, which never has to be physically measured, since it can be found using the Pythagorean theorem.

The visible distance to a horizon can be found with the Pythagorean theorem, given that the radius of the earth is 6,380 km. Inside a 100-meter-tall lighthouse, a night watchman or the coast guard may be interested in the distance a ship is from shore when seen at the horizon. This information can be readily found, since the horizon distance is perpendicular to the radius of the earth, forming a right triangle into the center of it, as shown below. The viewing distance inside the top of the lighthouse is then the solution to the equation $6380000^2 + b^2 = 6380100^2$, a value of over 35 km!

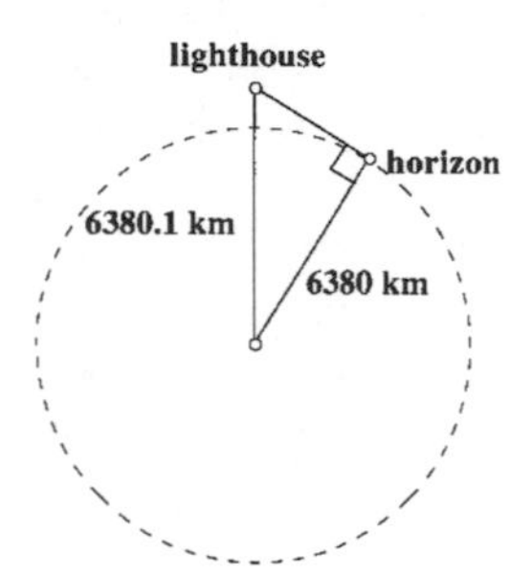

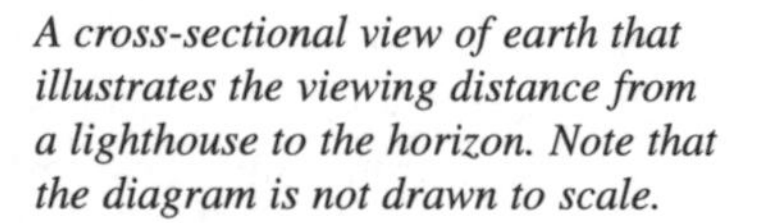
A cross-sectional view of earth that illustrates the viewing distance from a lighthouse to the horizon. Note that the diagram is not drawn to scale.

Extensions of the Pythagorean theorem provide distances in three or more dimensions. If a rectangular box has dimensions of length L, width W, and height H, the main diagonal has a length given by $d^2 = L^2 + W^2 + H^2$. Can a 42-inch-long umbrella be packed into a carton that is 40 inches long, 10 inches wide, and 10 inches high? According to the three-dimensional Pythagorean theorem, the diagonal is about 42.43 inches long. Yes, it should just barely fit. (See ***Vectors*** for applications in many dimensions.)

online sources for further exploration

Astronomy connections
<http://www.kyes-world.com/pythagor.htm>

Baseball and the Pythagorean Theorem
<http://www.geom.umn.edu/~demo5337/Group3/bball.html>
<http://www.pbs.org/wgbh/nova/proof/puzzle/baseball.html>

Construction
<http://www.geom.umn.edu/~hipp/app2.html>

Latitude and longitude
<http://daniel.calpoly.edu/~dfrc/Robin/Latitude/pythag.html>

Real-world applications
<http://www.geom.umn.edu/~hipp/rwapps.html>

When would I use the Pythagorean Theorem?
<http://forum.swarthmore.edu/dr.math/faq/faq.pythagorean.html>

▲▼▲

QUADRATIC FUNCTIONS

Quadratic functions take on the standard form $f(x) = ax^2 + bx + c$, and have graphs that are parabolas. Applications of quadratic functions commonly refer to maximizing or minimizing a quantity, because they will have a highest or lowest point at their vertex. For example, a business owner would be interested in the greatest profit his or her company can attain based on the sales of its products.

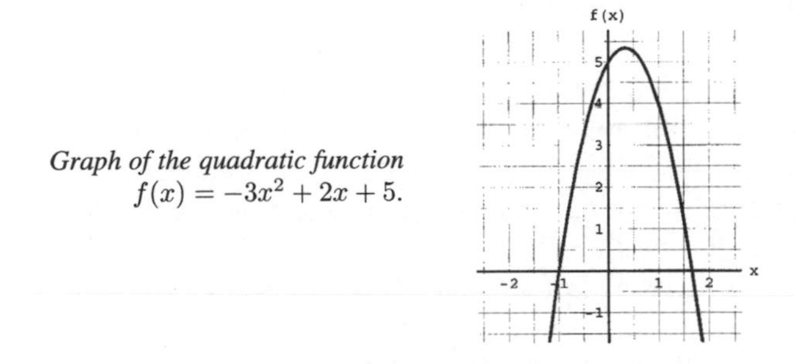

Graph of the quadratic function $f(x) = -3x^2 + 2x + 5$.

This maximum or minimum point can be found by rewriting the expression into vertex form through a process called *completing the square*. The vertex form of a quadratic function is $f(x) = a(x - h)^2 + k$, where (h, k) is the vertex. The following symbolic manipulation illustrates how the standard form $f(x) = ax^2 + bx + c$ can be manipulated into vertex form. Factor the leading coefficient, a, from the first two terms: $f(x) = a(x^2 + \frac{b}{a}x) + c$. Complete the square of the factored component, and then subtract that value so that nothing is added to the expression:

$$f(x) = a(x^2 + \tfrac{b}{a}x + \tfrac{b^2}{4a^2}) - \tfrac{ab^2}{4a^2} + c.$$

Rewrite the expanded trinomial as a perfect square and simplify:

$$f(x) = a(x + \frac{b}{2a})^2 - \frac{b^2}{4a} + c.$$

Compare this expression to the vertex form of a quadratic function and notice that the vertex can be represented as $(-\frac{b}{2a}, -\frac{b^2}{4a} + c)$. This coordinate will serve as a shortcut to find the highest point when $a < 0$, and lowest point when $a > 0$, in an application that uses quadratic functions.

Take, for instance, a business setting that sells sport memorabilia. The demand, and hence the price, for a baseball player's autographed ball may decline as more of them become available. Suppose that the price of an autographed ball, a, from a new hall-of-famer begins at $200 and declines by five cents, or $0.05, for every ball, x, sold. This relationship would be represented by the equation $a = 200 - 0.05x$. The revenue, r, obtained from selling the balls would be the product of the number of balls sold and the price for each ball, or $r = x(200 - 0.05x) = 200x - 0.05x^2$. The business owner will have to pay for general start-up costs such as hiring the baseball player to sign autographs and renting a place to sell the merchandise, as well as paying for the materials, such as the cost of each ball. Suppose the start-up costs are $1,300 and the business owner pays $1.25 for each new ball. Then the cost, c, that the business assumes in terms of the number of balls sold will be $c = 1.25b + 1300$.

The profit, p, is the difference between the revenue and cost, or $r - c$, which equals $(200x - 0.05x^2) - (1.25x + 1300)$, and simplifies to $p = -0.05x^2 + 199.75x - 1300$. In a quadratic function in the form of $f(x) = ax^2 + bx + c$, a maximum value will occur when $x = -\frac{b}{2a}$, since $a < 0$. In this case, a maximum profit will occur when approximately 1,997 balls are sold $(x = -\frac{199.75}{2(-0.05)} = 1997.5)$. In that case, a reasonable sale price of the "limited edition" ball should be around $200 - 0.05(1997) = \$100.15$. Although, to appease the human psyche, a more reasonable price might be twenty cents cheaper at $99.95 so that consumers feel like they are getting a deal by paying less than $100. A graph and table of values can also support this sale price as a means of producing almost a maximum possible profit.

The vertical height, h, of an object is determined by the quadratic equation $h = -0.5gt^2 + v_ot + h_o$, where g is the acceleration due to earth's gravity (9.8 m/sec^2), v_o is the initial vertical velocity, and h_o is the initial height of the object. Therefore an object with an initial vertical velocity of 45 meters per second, thrown at a height of 0.4 meters, can be modeled with the equation $h = -4.9t^2 + 45t + 0.4$. Engineers of fireworks can use this type of function so that the rockets explode at a time where optimal height offers safety as well as viewing pleasure.

This quadratic equation can also be used to measure the initial vertical velocity of an object thrown in the air, such as a ball, assuming that it reaches the ground with minimal air resistance. For example, if a ball thrown at a height of 1.45 meters is airborne for 3.84 seconds, then the values can be substituted into the equation $h = -0.5gt^2 + v_ot + h_o$ to solve for v_o. In this case, the height after 3.84 seconds is equal to 0 meters, because that is the amount of time it takes for the ball to reach the ground. The equation then becomes $0 = -0.5(9.8)(3.84)^2 + v_o(3.84) + 1.45$, which has a solution of v_o approximately equaling 18.4 meters per second. Substituting this value into the general function will also provide enough information to help you find the maximum height of your throw.

The equation $h = -0.5gt^2 + v_ot + h_o$ can be simplified to $h = -0.5gt^2 + h_o$ for objects in freefall because $v_o = 0$ when an object is dropped. Therefore if you plan to bungee-jump 200 meters off of a 250-meter-high bridge, then you should expect to be dropping for about 6.4 seconds. This value comes from substituting for the variables and solving the equation $50 = -0.5(9.8)t^2 + 250$. (Note that the ending position will be 50 meters above the ground, since the rope is only extending 200 meters.) This general equation could also be used to estimate heights and times for other objects that are released at high heights, such as the steep drops on some amusement park rides.

Horizontal distance, such as the distance traveled after slamming on the brakes in a car, can also be modeled with a quadratic function. In an effort to reconstruct a traffic accident, a law office could use the function $d = 0.02171v^2 + 0.03576v - 0.24529$ to determine how far a car could travel in feet, d, when breaking, or how fast it was moving in feet per second, v, before it started braking. The law office might also consider the average reaction time of 1.5 seconds upon seeing a hazardous condition. So the total stopping distance, t, can be modeled with the equation $t = 0.02171v^2 + 0.03576v - 0.24529 + 1.5v$, which simplifies to $t = 0.02171v^2 + 1.53576v - 0.24529$.

Area applications can also be modeled by quadratic functions, because area is represented in square units. For example, pizza prices depend on the amount of pizza received, which is examining its area. However, on a pizza menu, the sizes are revealed according to each pizza's diameter. If a 12-inch pie costs $12, a misconception would be to think that the 16-inch one should cost $16. A function to represent the price, p, of this type of pizza in terms of its diameter, d, is $p = 0.106\pi(\frac{d}{2})^2$, because it is a unit cost times the pizza's area. The value 0.106 is the price per square inch of pizza in dollars, assuming that the 12-inch pie for $12 will have the same unit-cost value as any other size pizza. Therefore a 16-inch pizza should cost $p = 0.106\pi(\frac{16}{2})^2 \approx \$21.31$. The restaurant, however, may decide to give a financial incentive for customers to purchase larger pies and reduce this price to somewhere near $20.

Devising and purchasing tin cans for food are applications of surface area that can be represented by a quadratic function. Since most tin cans are cylindrical, the surface area can be determined by finding the area of the rectangular lateral area and the sum of the two bases, as shown in the following figure. If the manufacturer determines the height of its cans to be 4 inches tall with a variable radius, then the amount of sheet metal in square inches, a, needed for each can would be $a = 8\pi r + 2\pi r^2$, where r is the radius of the can in inches.

If tin costs the manufacturer $0.003 per square inch, then the materials cost, c, to produce each case of twenty-four cans can be represented by the function $c = (24)(0.003)a = (24)(0.003)(8\pi r + 2\pi r^2)$, which simplifies to $c \approx 1.81r +$

$0.45r^2$. Therefore a case of cans with a radius equal to 1 inch would cost about \$2.26 to produce, and a case of cans with a radius equal to 2 inches would cost about \$5.42. (See ***Surface Area***.)

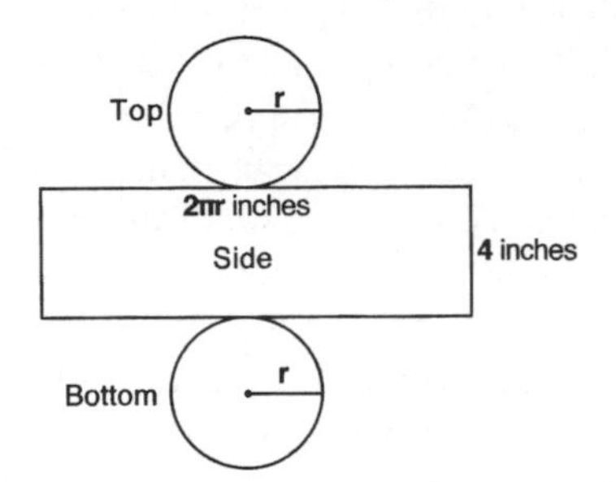

Net of a tin can that is used to determine the amount of material needed in manufacturing production.

Temperature usually changes according to changes in elevation. In fact, the boiling point of water in degrees Celsius, b, can approximate the elevation above sea level in meters, e, according to the equation $e = 1000(100 - b)^2 + 580(100 - b)$. This means that water will boil at 100° Celsius near sea level, and closer to 99° Celsius at about one mile in altitude, such as near Denver, Colorado.

online sources for further exploration

Braking and stopping distances compared with speed
<http://www.exploratorium.edu/cycling/brakes2.html>
<http://www.scottsdalelaw.com/shepston/braking.html>

The ejection seat and parabolic paths
<http://daniel.calpoly.edu/~dfrc/Robin/Eject/eject.html>

Fluid flow
<http://www.imacc.org/standards/ex15.html>

Minimum surface area of a can
<http://jwilson.coe.uga.edu/emt725/MinSurf/Minimum.Surface.Area.html>

Optimization and analysis using quadratic functions
<http://www.wake.tec.nc.us/math/Chimp/Unit3/QUADRT_S.html>

Profit
<http://users.aber.ac.uk/matacc2/ma12610/mich00b/node3.html>

Projectile motion simulations
<http://library.thinkquest.org/2779/Balloon.html>
<http://www.explorescience.com/activities/Activity_page.cfm?ActivityID=19>
<http://www.phys.virginia.edu/classes/109N/more_stuff/Applets/ProjectileMotion/jarapplet.html>

The way things fall
<http://www-spof.gsfc.nasa.gov/stargaze/Sfall.htm>

QUADRILATERALS

Four-sided plane figures are called *quadrilaterals*. Quadrilaterals can be convex or concave. The wing structure of the B-117A bomber is in the shape of a convex quadrilateral (white outline on the illustration below). Special types of quadrilaterals such as rectangles and squares are used for warning signs and flags. The illustration shows these common kinds of quadrilaterals:

- a convex quadrilateral superimposed on the wing structure of a F-117A *Nighthawk*;
- a square traffic sign;
- a rectangular flag;
- an isosceles trapezoid superimposed on the bottom section of the John Hancock Building in Chicago;
- a kite;
- parallelogram faces of a Moissanitc-9R CSi crystal structure;
- diamonds (rhombi) on a playing card;
- a city lot in the shape of a trapezoid.

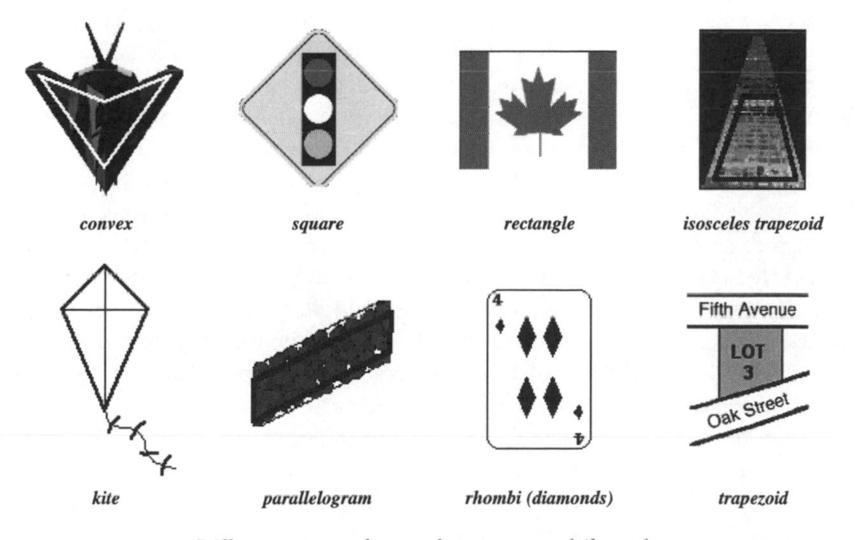

Different types and uses of common quadrilaterals.

The types of quadrilaterals differ in the number of pairs of parallel sides, size of angles, and length and direction of diagonals. The parallelogram has parallel opposite sides. As a result, opposite angles are congruent, and opposite sides are congruent. If a parallelogram has all four sides of the same length, then it is a *rhombus*. This results in the diagonals of a rhombus being perpendicular. A parallelogram that has at least one right angle is a rectangle. It must have all right angles and diagonals that are the same length. A square is simultaneously a rectangle and a rhombus. Hence it has only right angles, its diagonals are congruent and perpendicular, and all four sides are of equal length.

There are many uses of parallelograms in carpentry. A four foot by six foot window frame is made by connecting four-foot pieces to the six-foot pieces so that opposite sides of the frame are equal in length. The resulting figure is a parallelogram. Even though the corners are securely connected, it is likely to shear so that the angles are not 90 degrees. Even a slight deviation may mean that a rectangular window will not fit into the frame. To square the frame, the carpenter measures the diagonals. The parallelogram frame will be rectangular only when the diagonals have the same length. When the frame is in the wall opening, the carpenter will use shims, small pieces of wood, to adjust the fit of the frame until the diagonals are the same length. Then the carpenter can be sure that the rectangular window will fit into the frame.

Doors are usually shaped like rectangles so that they can seal better at their hinge, or at their edges. An entire side of a door can be well connected to a set of hinges, as well as allow a person to easily walk through. If the door were shaped as an oval it would be primarily for design purposes, since the door would likely be less durable and more expensive. Only a small section could be attached to maybe one hinge, affecting its ability to stay well connected to the house. In addition, walking through the doorway would be more challenging, because less floor space would be available as compared to the flat edge of a rectangle. Sometimes doorways are rectangular and have an arch at the top, which is an architectural style seen in many cultures. It is built principally for design or historical significance and is usually more difficult and expensive to construct.

Floor tiles in the shape of the same quadrilateral will always fit perfectly, because the sum of any quadrilateral's angles is 360 degrees, the same measure of degrees in a circle. In order to tessellate a plane, all of the objects must connect perfectly without any gaps or overlaps—what you would expect of tiles in a bathroom or kitchen. At the point where multiple tiles intersect, their interior angles must equal 360 degrees so that they fit neatly around a common center point. If the different angles of a quadrilateral are used around an intersection of four quadrilaterals, they will always tessellate perfectly.

Trisection of a long piece of lumber into thinner strips requires that guide lines be set up for the ripsaw. A carpenter can take a 12-inch ruler and rotate it so that its opposite ends are at the edges of the lumber. After marking the board at the 4 and 8 inch positions at one end, the carpenter slides the ruler down the board and marks the 4 and 8 inch positions at the other. Corresponding marks are used to draw long lines down the board as guides for the saw. This works, because the marks at 4 and 8 inches provide vertices of a parallelogram.

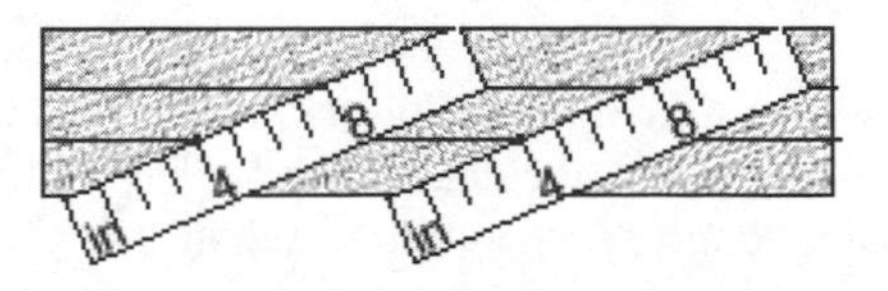

Using an inch ruler to divide a board lengthwise into three equal sections.

A *pantograph* is used to produce a scale drawing. The picture below of a pantograph shows that the four pivot points that connect the pieces of wood form a parallelogram. The ratio of lengths of sides controls the scale of magnification. Pantographs that handle three dimensions are used to trace solid models of bolts, car fenders, or teeth. The pantograph records the three-dimensional coordinates for the surface of the object. Milling machines use the database of coordinates to shape a block of metal, plastic, or carbon composite into a high-precision copy of the original object.

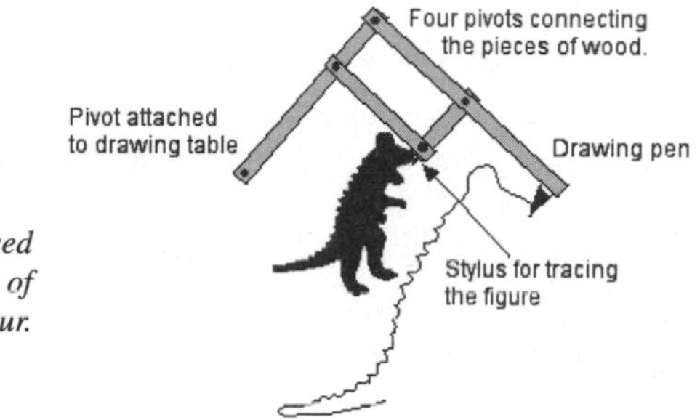

A pantograph being used to magnify a picture of a dinosaur.

The *parallelogram law* is used in physics to determine the net result of two forces. The vectors $\vec{a} = (3,3)$ and $\vec{b} = (7,-1)$ are shown on the figure below as arrows starting at the origin and ending at the respective coordinates. The parallelogram law indicates that the *resultant vector* is found by completing the parallelogram defined by the vectors. The diagonal from the origin is the desired vector. This corresponds to the point that would be found by the addition of coordinates: $(3,3) + (7,-1) = (10,2)$. (See ***Vectors***.)

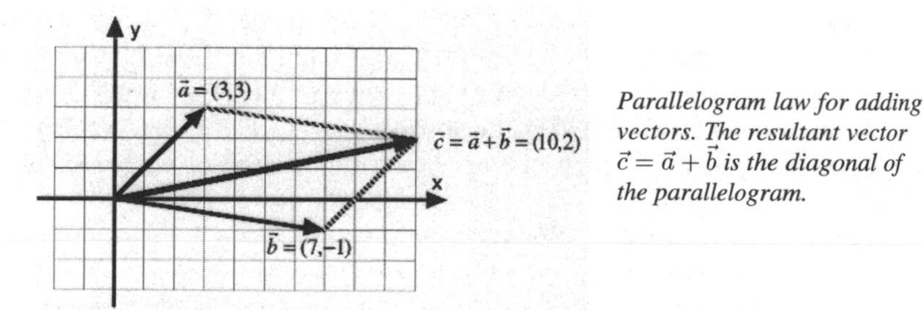

Parallelogram law for adding vectors. The resultant vector $\vec{c} = \vec{a} + \vec{b}$ is the diagonal of the parallelogram.

Because of the many uses of quadrilaterals, students around the world are expected to know formulas for the area and perimeter of most common quadrilaterals. In addition, they must also know the volume formulas for the three-dimensional analogs of some quadrilaterals such as the cube and rectangular solid. Formulas for the multidimensional parallelepipeds are expressed as determinants of matrices formed from the vectors. For example, the area of the parallelogram illustrated above can be computed from the determinant of $\begin{bmatrix} 3 & 7 \\ 3 & -1 \end{bmatrix}$ $= 3(-1) - 7(3) = -24$. The area is 24. (See ***Matrices***.) Extensions to more dimensions provide measures of strength of association of variables in multivariate statistics.

online sources for future exploration

Demonstration of a pantograph
<http://www.ies.co.jp/math/java/geo/panta/panta.html>

Floor plans
<http://www.homebuyerpubs.com/foorplans/floorplans.htm>
<http://www.dldesigngroup.com/plans.html>
<http://ecep.louisiana.edu/ecep/math/n/n.htm>
<http://www.tnloghomes.com/homeplans/index.shtml>

Home decorating
<http://www.learner.org/exhibits/dailymath/decorating.html>

Maximize the area of a rectangular field with fixed perimeter
<http://home.netvigator.com/~wingkei9/javagsp/maxarea.html>

Surveying
<http://www.math.bcit.ca/examples/ary_17_2/ary_17_2.htm>

Tessellation of quadrilaterals
<http://library.thinkquest.org/16661/simple.of.non-regular.polygons/quadrilaterals.html>

RATES

A *rate* describes change that is dependent on a variable, such as the change of temperature in a month or the change in price of an item based on the quantity sold. The concept of rate is studied throughout mathematics in different forms. It can usually be identified in an expression by the word "per," such as in "two dollars per gallon"; or "for each," as "one-half unit of credit earned towards graduation for each required course completed"; or "for every," as in "six points for every touchdown scored."

Rates are commonly associated with the amount of distance traveled, d, in the equation $d = rt$, where r is the rate of an object and t is the amount of time traveled. In this case, the rate would be expressed in units related to speed, such as meters per second or miles per hour. Rate can also be used in contexts of production levels for a given time period. For example, two hundred bushels of corn are processed by the manufacturing plant each day, or twenty copies of the newspaper are sold each hour at the newsstand. Rate is also associated with accumulating or acquiring something, such as rainfall or a salary. For instance, the thunderstorm is producing rain at a rate of one inch per hour. At work, an employee would use a rate to describe his hourly wage by saying that he earns \$8 per hour for delivering pizzas.

The examples described thus far represent rates as values associated with time. Rates can also be stated in terms of quantities produced or achieved. For example, the delivery boy receives five cents for each newspaper he drops off each morning. In addition, Mrs. Newsome's first-grade class receives twenty minutes of extra playtime for every one hundred behavior points earned. In a securities exchange, a rate can be used to illustrate a fair trade, such as in stock or currency values. For instance, the exchange price of Big Hit Co. today ended at $48.5 per share. When traveling to Mexico, you would expect to receive an exchange rate of about 9.3 pesos for every U.S. dollar.

Rates can also be used to describe changes in an environment or physical setting. For example, two hundred additional employees are needed for every 8 percent increase in demand for the company's products. In terms of temperature conversion, there is a change of 1.8° Fahrenheit for every degree Celsius. When driving along a mountain terrain, a road sign that mentions a 5 percent grade means that there is a change in elevation of five vertical feet for every one hundred horizontal feet.

Many scientific, engineering, and human measures are rates. Density is a weight-per-volume measure such as pounds per cubic foot or grams per cubic centimeter. Sound frequencies, such as those associated with musical notes, are expressed as rates in cycles per second. Air pressure, such as tire pressure, is expressed as pounds per square inch. The wealth of countries is compared as the rate of Gross National Product (GNP) per capita. In 1997, Mexico had GNP per capita of $8,110; Canada had a GNP per capita of $21,750. States can be compared by population density: the number of people per square mile. Comparisons may be dramatic. For example, New Jersey has 1,100 people per square mile, while Wyoming has 4.7.

Comparison shopping requires rates. If an eight-ounce can of corn sells for 98 cents, the unit cost is 98/8 = 12.25 cents per ounce. A ten-ounce can that sells for $1.02 would have a unit cost of 102/10 = 10.20 cents per ounce. The larger can is the better deal, because it provides the lower unit cost.

Rate, in mathematics courses through algebra, is often presented as having a constant value. When you read about the speed of an object or a person's work wages, it is assumed that there will not be any change in these values. In such cases, the rate can be represented as the slope of a linear function that describes a total amount. For example, if you are earning $8 per hour for delivering pizzas, and always earn wages at that rate, then your total earnings, e, in terms of the number of hours you have worked, h, can be represented by the equation $e = 8h$. Notice that the hourly rate is the same as the slope of the linear function.

Suppose you wanted to make copies for a class presentation at the local copy shop. If the machine charges 10 cents per copy, then the total amount of money, m, that you would need would depend on the number of copies, c, you make. Since 0.10 is the rate in dollars, the equation $m = 0.10c$ would help you determine the amount of money you would need, or the number of copies you could make with a certain amount of money. For example, if you had $4.30 in your pocket you could make forty-three copies, since the solution to $4.30 = 0.10c$ is $c = 43$.

Realistically, rates are often variable, meaning that they change. A car on the highway will not always travel 55 miles per hour because of varying road conditions. If traffic is heavy due to rush hour or an accident, the car will likely slow down at times. Therefore the average rate is sometimes stated in reports. The average (mean) rate can be calculated by finding the slope between beginning and ending points on the graph that represents a total amount. For example, if a car is traveling at a constant speed of 55 miles per hour, then the total distance traveled as a function of time would be a linear function with a slope of 55, as shown below.

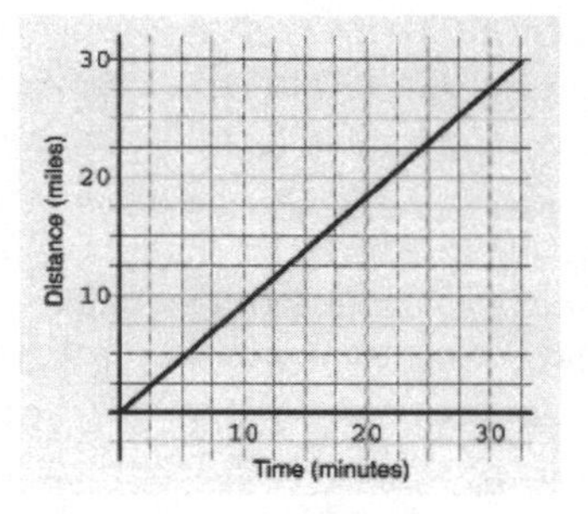

The relationship between the distance an automobile travels compared to its time when it travels at a constant rate of 55 miles per hour.

However, if the car varies its speed, the total distance function will now look like a curve that does not have a constant slope. If a car travels for three hours on the highway, the average speed can be determined by finding the slope of the line that time equals 0 and 3 hours. According to the slope between the endpoints in the graph in the figure below, the average speed during the three hours is 49 miles per hour, since the change in distance was 147 miles over three hours.

A distance versus time graph of an automobile with varying rates during a three-hour time period. The slope of the dotted line is the automobile's average speed during that time period.

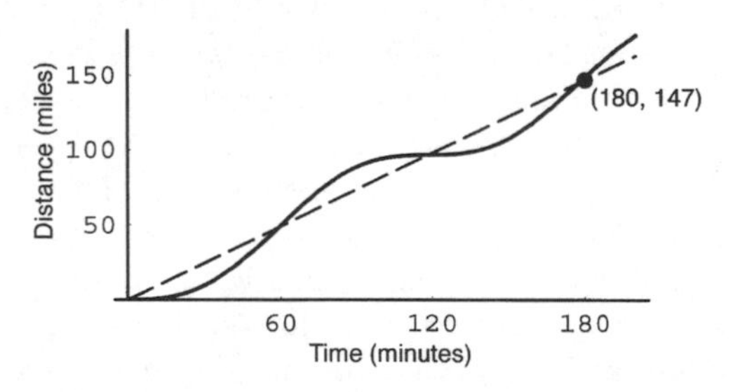

Some highway systems in the United States give a timed ticket for automobiles once they enter on the toll road so that they can pay the fee at the end of their route instead of having to pay along the way. Upon exiting the highway and paying the toll, the highway patrol system can determine the average speed of the vehicles during their travel by dividing the change in distance between the tollbooths and the change in time from the initial starting point to the ending point. For example, suppose you enter tollbooth 3 at mile-marker 27 at 12:34 PM. If you leave the highway at tollbooth 17 at mile-marker 136 at 1:57 PM, you could actually receive a speeding ticket without having been tracked by a radar speed-detection device! In this case, the change in distance between tollbooths 3 and 17 is 109 miles, and the change in time between 12:34 PM and 1:57 PM is 1 hour and 23 minutes, or approximately 1.38 hours. Therefore the average speed of the car is about 79 miles per hour, much faster than the speed limit! The mean-value theorem in calculus implies that a car constantly in motion with this average speed will have traveled at that rate at least one time during the journey, even if undetected by radar. The graph below describes the position of the car for its time on the highway. The dotted line represents the average rate of 79 miles per hour. The three times that the car was traveling at 79 miles per hour are indicated with the word "speeding." Note that there are many other times that the car was speeding more than 79 miles per hour. The mean value theorem from calculus only tells that there is at least one time that the car had to be going the average rate of 79 miles per hour.

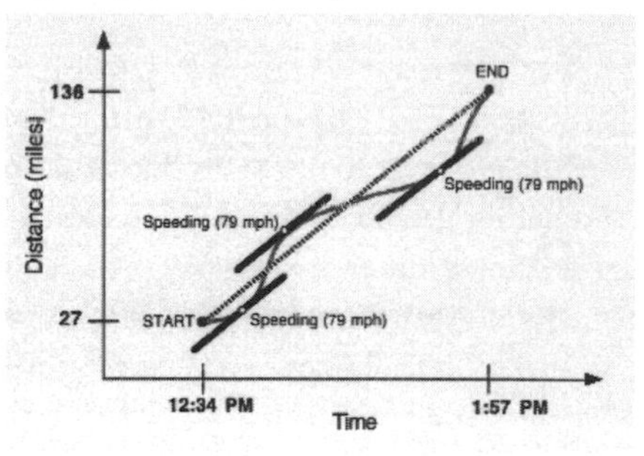

The slope of the dotted line represents the average speed of the car from 12:34 PM to 1:57 PM, which is 79 miles per hour. The automobile has this rate at three other locations in this interval based on the equivalent slopes of the small thick lines (at the points denoted speeding).

In addition to tracking speeding drivers, the time-stamping method is also helpful in determining the average speed of truck drivers, who need to take breaks from the road so as not to fall asleep behind the wheel. Consequently, the average speed of semi-trucks should be lower than other automobiles to account for the rest time.

The average rate associated with the slope on an interval is also an arithmetic mean. Sometimes average speed can use other forms of the word *average*. On a racetrack, car speeds are determined by finding the average of the lap rates. This value is different from the average speed determined by the slope of a position function, which is the same as the total distance divided by the total time traveled. For example, suppose a race car circles a two-mile lap five times, with lap times of 46, 48, 47, 45, and 49 seconds. In this case, the lap speeds would be 2/46, 2/48, 2/47, 2/45, and 2/49 miles per second. The recorded average speed would be the average of these rates,

$$\frac{2/46+2/48+2/47+2/45+2/49}{5} = 4060879/95344200 \text{ miles per second},$$

which is approximately 153.33 miles per hour. If an arithmetic mean were used to determine this rate, then the total distance traveled, ten miles, would be divided by the total time taken for five laps, 235 seconds. This value of 10/235 miles per second, or approximately 153.19 miles per hour, may be a more accurate representation of the average speed of the car. Since lap time is more easily and commonly tracked continuously throughout the race, the average lap speed is used instead of the average rate.

Besides finding the average rate as a means to describe varying speeds, it is possible to determine the instantaneous rate of an object using differential calculus. If a total amount, such as distance or production levels, can be described as a function, then the rate at any moment can be determined by finding the derivative of that function. Instead of finding the slope at the endpoints of an interval, a derivative is the slope of a line tangent to a curve at a particular point.

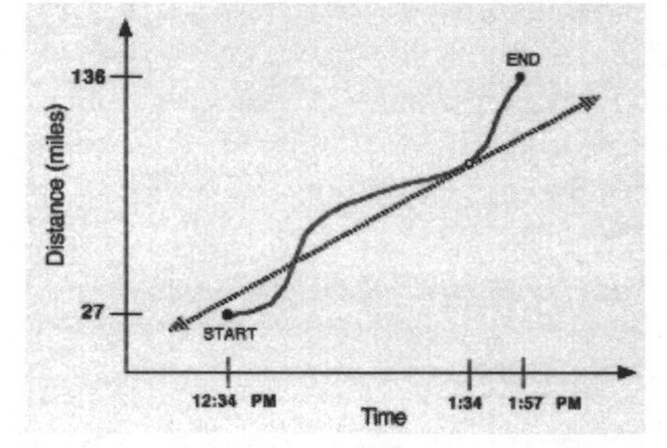

The slope of the dotted line that is tangent to the curve is the automobile's instantaneous speed of 70 miles per hour at 1:34 PM.

The slope of the tangent line will describe the speed of the car at a specific moment in time. For example, in the above figure, a tangent line with a slope of 70 miles per hour is drawn on the curve at 1:34 PM, illustrating the speed of the car at that moment.

In addition to automobile travel, the motion of falling objects shows variable rates. Since the earth pulls objects at a rate of 9.8 meters per second squared, falling objects are constantly accelerating. The position of a penny dropped off of a 400-meter-tall skyscraper can be represented by the function $h = -4.9t^2 + 400$, where h is the height of the penny above the ground in meters, and t is the time in seconds the penny is airborne. This function is a parabola. It will not have a constant slope, which means that the penny will not fall at the same rate towards the ground. However, the slope of the line tangent to the curve at any time, or the instantaneous rate, can be predicted by the derivative of this function, which is $h' = -9.8t$. This means that the penny will be falling at a rate of 9.8 meters per second after one second, 19.6 meters per second after two seconds, and so on. According to the position function, $h = -4.9t^2 + 400$, the penny will reach the ground at approximately $t = 9$ seconds, where h is equal to 0. According to the derivative of the position function, the velocity of the penny by the time it hit the ground would be $h' = -9.8(9) = -88.2$ meters per second, fast enough to fall straight through a person's body. Hence, you are not likely to be permitted to drop objects from tall buildings!

Human workforce productivity can have varying rates. In a factory, the workers may be less productive in the early morning because they are tired, and then reach an optimal work rate later in the morning when they are more awake. Later in the afternoon, they may become less productive again due to fatigue or boredom. Understanding the varying working rates of employees may help management determine an optimal time to take a break or to change work shifts. Knowing the change in work rates would provide information to make smart decisions on behalf of the safety of the employees, as well as to support the economic benefits of the company.

In consumer sales, the profit from a business is often dependent on the number of products sold. An ideal production level would be to determine the moment when the change of profit, or the rate at which profit is changed, begins to level off to zero. According to supply-and-demand principles, the company would like to produce the appropriate amount of products in order to meet consumer demand, but not end up with a surplus in inventory. If too many goods are produced the rate of profit declines, because the company would lose money on excess inventory. For example, suppose the price per cup of lemonade, l, depended on the number of cups, n, purchased according to the equation $l = 2.00 - 0.01n$. This equation suggests that the price of a cup of lemonade would be $2.00 if none were sold, but the price will decline by one penny for every cup sold. The revenue, r, obtained for selling lemonade would be the product of the price per cup and the number of cups purchased. Therefore the total revenue would be equivalent to $r = lc = (2.00 - 0.01n)n = 2.00n - 0.01n^2$. The cost to make the lemonade depends on start-up expenses and the quantity of lemonade sold. If the lemonade stand costs $12.00 to set up and each cup costs $0.14 to produce, then the cost, c, for the company to make lemonade can be represented by the equation $c = 0.14n + 12.00$. The profit, p, obtained by selling lemonade is the difference between the revenue and costs, which is $p = r - c = (2.00n - 0.01n^2) - (0.14n + 12.00) = -0.01n^2 + 1.86n - 12.00$. The graph of the profit function is a parabola, illustrating that the rate of profit changes, because the graph is nonlinear. Notice that the maximum profit of the function occurs when ninety-three cups are sold—the moment when the rate of profit is equal to zero or where the slope of tangent line equals zero, as shown in the figure below.

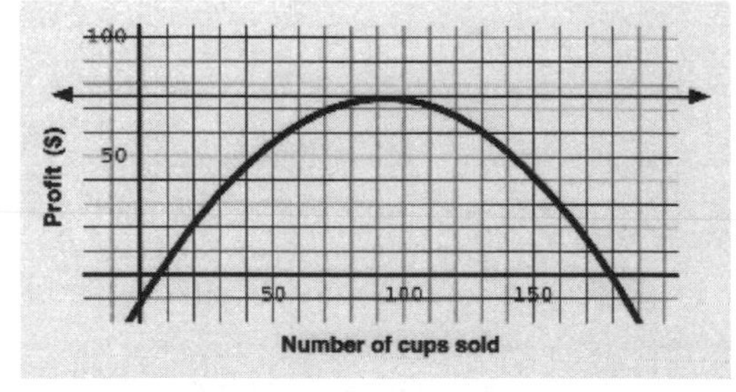

A maximum value of a quadratic function can be found by locating the horizontal tangent line.

online sources for further exploration

Designing a speedometer
<http://barzilai.org/archive/lc/speedometer.html>

Distance between two ships
<http://www.nadn.navy.mil/MathDept/cdp/relatedrates/rates.html>

Fair division activities
<http://www.colorado.edu/education/DMP/activities/fair_division/>

Gross national product data
<http://www.economagic.com/em-cgi/data.exe/fedstl/gnp+1>

Motion
<http://www.mste.uiuc.edu/murphy/MovingMan/MovingMan.html>
<http://webphysics.ph.msstate.edu/jc/library/2-6/index.html>

Motion of a piston
<http://www.math.bcit.ca/examples/ary_16_5/ary_16_5.htm>

Occupational health and safety
<http://www.math.bcit.ca/examples/ary_12_1/ary_12_1.htm>

Pipe flow
<http://www.math.bcit.ca/examples/ary_8_3/ary_8_3.htm>

Roofing
<http://www.professionalroofing.net/past/march00/qa.asp>

Slope use permit
<http://www.ci.larkspur.ca.us/3025.html>

Stressed out: slope as a rate of change
<http://math.rice.edu/~lanius/Algebra/stress.html>

Universal currency converter
<http://www.xe.net/ucc/>
<http://www.wildnetafrica.com/currencyframe.html>

Exchange rates for world currencies
<http://www.x-rates.com/>

RATIO

A *ratio* is a quotient of two numbers. One of the most famous ratios in mathematics is $\pi \approx 3.14159$, the ratio of circumference of a circle to the diameter. A ratio is different from a *rate*, in that the units for the numerator and denominator in a ratio are the same. A ratio does not have any units of measurement, unlike rates that have units such as miles per hour or dollars per pound. Some examples of ratios that are given here are really rates, but it is common practice in particular occupations and sciences to call them ratios.

There are many statistics dealing with money that are ratios. The federal government maintains the Consumer Price Index (CPI) and Cost of Living Index (COL). The CPI is the ratio of costs of common items in the current year to the costs of the same items during 1982–1984. The costs are usually expressed as a multiple of 100, so that the number represents the current cost of purchasing goods and services that would have cost $100 during 1982–1984. In 2001, the

CPI was 177. This means that the ratio of costs for goods and services in 2000 was 1.77 times as high as the costs for the same items during 1982–1984. The COL is computed for almost two hundred metropolitan areas. It reflects the ratio of costs of goods and services in a specific area to the average for the country as a whole. The COL is expressed as a percent. At the beginning of 2001, the COL for San Francisco was 179.8, and for Houston, 95.8. Those ratios mean that it costs 1.798 times the U.S. average to live in San Francisco, but 95.8 percent of the national average to live in Houston. Stockbrokers use the price-earnings ratio (P/E) as a way of evaluating stocks. This ratio is defined as the market value per share divided by the earnings per share. If a company has stock valued at $40 per share, and has earned a net of $2 per share over the last year, the P/E ratio for the company would be $40/$2 = 20. Most stocks traded on the major exchanges have P/E ratios between 15 and 25. Riskier stocks that have potential for rapid growth are likely to have P/E ratios above 25, if any at all. (If a company has not produced any earnings, then its P/E ratio is reported as 0.) In these circumstances, people invest their money in companies that they think will have low P/E ratios or a high demand in the future. The P/E ratio is only one of many ratios routinely reported for stocks.

Percents are based on ratios. If a taxpayer pays $3,000 on an income of $20,000, then the tax ratio is 3000/20000 = 15 percent. The federal government refers to this as a tax rate. If an investment of $500 is now worth $550, the percent increase is the ratio of absolute change to starting value, or 50/500 = 10 percent. When you specify how long it took for this increase, you express the change as a percent per year, or interest rate.

Test scores are often reported as *percentile ranks*. A student with a percentile rank of 60 on a college placement test achieved a score that was equal to or higher than the scores of 60 percent of the students taking the test. Therefore the percentile rank is a ratio of counts of students.

Body mass index (BMI) is the quotient of your weight in kilograms divided by your squared height, where height is measured in meters. Although this measure is a rate (kilograms/m^2), the units are not reported and are not used in computations of other health measures. BMI values from 20 to 25 are associated with the lowest health risks; values above 30 are associated with the highest. Weight-to-hip ratio (WHR) is a true ratio that indicates whether an adult carries weight around the waist or hips. Weight carried around the middle (higher ratio) is associated with more health problems. Many ratios developed for human physiology are applied to other animals. The ratio of an animal's surface area to its volume measures how much energy the animal has to produce to counter the heat lost through the skin. (See ***Inverse (Multiplicative)*** for an additional explanation.) A mouse has a surface-area-to-body ratio that is about ten times that of a human, so the mouse has to eat almost all day long to maintain its body heat, while a human does quite well with three meals per day.

The modern musical scale is based on a consistent ratio of frequencies from note to next note for the twelve notes of an octave. Middle C-sharp (275 cycles per second [cps]) is about 1.059 times middle C (260 cps); D (292 cps) is about 1.059 times C-sharp. This sequence continues to high C (520 cps), which is twice the frequency of middle C. Pythagoras (as later corrected by Galileo) tried to identify simple integer ratios for what would correspond to the white keys of a piano. The Pythagorean scale uses 9/8 for the ratio that would compute D from middle C (9/8 of 260 = 292.5).

Time signatures found at the beginning of a piece of music look like fractions without the fraction bar. They represent beat patterns for the measure. The notation $\substack{3\\4}$ means that there are three beats per measure, and a quarter note (1/4) receives one beat. This sets the ratio of note values to measures for the piece of music.

Almost all mechanical objects in your house use gears. A videotape machine uses gears to control tape motion. Windup and pendulum clocks use gears. Bicycles use gears. Gear ratios tell how rotational motion changes when you connect gears with different numbers of teeth. When a small gear with forty-seven teeth connects with a larger gear with seventy teeth, the gear ratio is $\frac{47}{60}$. The gear ratio can be used to compute how many times the larger gear will rotate compared to the smaller gear. (See ***Rotations*** for additional information about gears.)

Ratios that express mixtures are often written with a colon. When a gardening expert recommends two parts of sand, five parts of potting soil, and one part perlite for the soil mixture in a window box, the ratios can be written in one expression as 2:5:1. A fertilizer that is labeled as 25-5-5 represents the percents of nitrogen, phosphate, and potassium. The high ratio of nitrogen to the other substances means that this fertilizer is probably for the quick development of lawns, which need nitrogen. A fertilizer with a lower ratio of nitrogen like 10-20-20 would be good for a garden. Directions for recipes and household products are often given in ratios of parts. A wedding punch is two parts orange juice, two parts lemonade, one part pineapple juice, and one part grapefruit juice. The juices are in ratio of 2:2:1:1. (See ***Proportions*** and ***Similarity*** for additional applications of ratios in this form.)

The *golden ratio* or *golden section* is based on a rectangle that can be split into a square and a smaller rectangle that is similar to the original rectangle. The ratio of length to width of the original rectangle is $\frac{1+\sqrt{5}}{2} \approx 1.61803$. The ancient Greeks believed that this rectangle embodied the most satisfying proportions. The Parthenon in Athens fits the golden ratio. Some sociologists have argued that people who have certain facial features close to the golden ratio are judged by others as being more beautiful or handsome. The golden ratio expresses many patterns in plant and animal structures. (See ***Fibonacci Sequence*** for more information about applications of the golden ratio.)

Measures in science and engineering that produce extremely large numbers are simplified by ratio measures. In aviation, the Mach number indicates the ratio of the plane's speed to the speed of sound. Mach 1 is a critical value for airplanes. Below the speed of sound, a plane pushes air aside like a boat traveling through water. But when the plane hits the speed of sound, the airwaves can't move out of the way of the plane. The build up at the front of the plane causes a shock wave that creates stress on the plane and is often audible to people on the ground as a "sonic boom." The speed of sound varies according to temperature and other factors. It is about 762 miles per hour at sea level, and about 664 miles per hour at 35,000 feet. A jet traveling at 1,400 miles per hour 35,000 feet above sea level would be traveling at 1400/664 $\approx$ Mach 2.1. A jet-propelled wheeled vehicle achieved Mach 1.02 on the Bonneville Salt Flats on a day when the speed of sound was 748 mph. Its speed was 763 miles per hour.

Astronomers measure solar-system distances with a ratio measure called an *astronomical unit* (AU). An AU of 1 represents the average distance of the earth to the sun, about 14,960,000,000 kilometers. For even larger distances than the solar system (which is about 80 AU in diameter), astronomers use ratio measures based on light years. One light year is the distance traveled by light in one year (about 9.46×10^{17} cm). Our galaxy is about 100,000 light years in diameter. Parsecs (3.26 light years), kiloparsecs (1,000 parsecs), and megaparsecs (1 million parsec) are used to measure distances across many galaxies.

Trigonometric ratios are used to find unusual or inaccessible heights and lengths. By measuring angles and shorter distances, an engineer can calculate the height of skyscrapers by creating diagrams with right triangles and using these ratios. (See ***Triangle Trigonometry*** for an explanation.)

Scale models use ratios to indicate how the lengths of an object compare to corresponding measures in the model. A 1:29 scale-model train would be large enough for children to ride outdoors on top of the cars. It would be 1/29th of the size of a real train. An HO-gauge tabletop train is at a scale of about 1:87. An 8.64-inch model of an 18-foot-long automobile (216 inches) would be at the scale of 1:25. Scale models can also help provide information to calculate unknown information, such as the mass of a dinosaur. (See ***Similarity***.) Although the design of buildings, cars, toasters, and furniture may involve drawings and models that are smaller than the final version, scale models that are larger than real life are important in many fields. Manufacturers of computer chips make scale drawings much larger than the actual chip to show the packed circuitry. Medical researchers make large-scale models of viruses and cell structures to determine how shapes affect resistance to disease.

The *fundamental law of similarity* uses scaling to indicate how surface area and volume of the model relate to the actual object. If k is the ratio of a length in the object to the corresponding length in the model, k^2 is the ratio of surface areas, and k^3 is the ratio of volumes. This law explains the limits on human and animal growth. If a six-foot-tall, 180-pound human were to double in size so that his relative proportions were maintained, he would be twelve feet tall, but his volume, and hence his weight, would be eight times as much. The giant's weight would be 1,440 pounds—which couldn't be supported by human bone structures. (See ***Proportions*** for an alternate explanation.)

online sources for further exploration

Consumer price index (CPI)
<http://stats.bls.gov/cpihome.htm>

The P/E ratio and other stock ratios are discussed at the Motley Fool page
<http://www.fool.com/School/EarningsBasedValuations.htm>

Musical scales
<http://hyperphysics.phy-astr.gsu.edu/hbase/music/pythag.html>

FAA instructions on making a scale drawing of an F-16
<http://www.faa.gov/education/resource/f16draw.htm>

Compute gear ratios for a bicycle
<http://home.i1.net/~dwolfe/gerz/howto1.html>
<http://www.panix.com/~jbarrm/cycal/cycal.30f.html>

U. S. Census Bureau QuickFacts on States (Rates and Ratios)
<http://quickfacts.census.gov/qfd/>

Body-mass calculator
<http://cc.ysu.edu/~doug/hwp.cgi>
<http://www.jsc.nasa.gov/bu2/inflateCPI.html>

Cooking by numbers
<http://www.learner.org/exhibits/dailymath/cooking.html>

Density lab
<http://www.explorescience.com/activities/activity_page.cfm?activityID=29>
<http://www.panix.com/~jbarrm/cycal/cycal.30f.html>

How to compute baseball standings
<http://www.math.toronto.edu/mathnet/questionCorner/baseball.html>

Scale models
<http://www.faa.gov/education/resource/f16draw.htm>
<http://www.pbs.org/wgbh/nova/pyramid/geometry/model.html>
<http://www.americanmodels.com/sscale.html>

Screen ratios
<http://www.premierstudios.com/ratio.html>

Tuning in
<http://www.bced.gov.bc.ca/careers/aa/lessons/aom15.htm>

REFLECTIONS

A *reflection* is a transformation that produces an image of equal size by flipping an object over a line. For example, you will see a reflection of yourself when you look in the mirror. Your size in the mirror will be the same as your actual size, but all of your features will be reversed. So if your hair is parted to the left, it will appear to be parted to the right in a reflection. Using two mirrors can create double reflections, allowing someone such as a hair stylist to show you the back of your head after a haircut while you look straight ahead.

Reflections of objects are naturally visible in water. If you walk up to a pond on a still, sunny day, you will see an image of yourself on the surface of the water. In the picture below, buildings and boats on a Holland canal are reflected in the surface of the canal. The reflection is so good that when you turn the picture upside down, it looks almost the same.

Boats and buildings reflected in a canal in Holland. Source: Adobe Stock Photos.

Reflections are sometimes used to create illusions or expand the size of an object. Many restaurants have large mirrors on one wall so that the room will appear twice as large. In an amusement park, a house of mirrors creates multiple images of anyone walking through, making it difficult to determine the correct pathway to the exit. Another example of using reflections to replicate an object is to create designs with a kaleidoscope. A kaleidoscope is a cylindrical toy that creates colorful patterns by using tiny objects situated at its base and in between two intersecting mirrors. The reflections at the base repeat themselves as a function of the angle n between the mirrors. Since there are 360 degrees in a circle, then there will be $\frac{360}{n}$ repetitions of the object caused by reflections. Each time the kaleidoscope rotates, the tiny objects inside it move around and consequently change the symmetrical pattern one sees when looking through the cylinder.

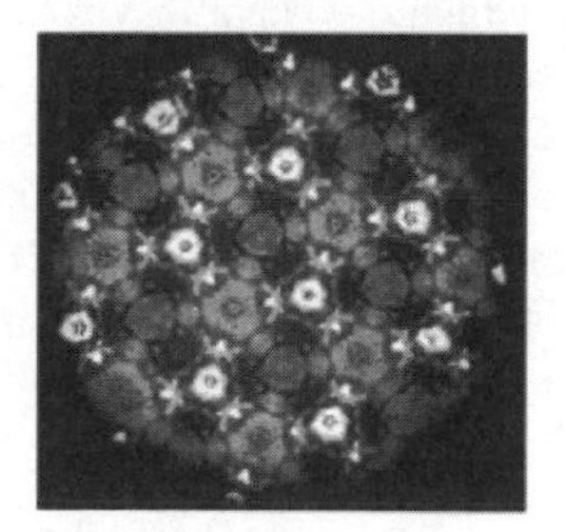

A kaleidoscope uses mirrors to produce multiple reflections and create colorful patterns. Photo by Dror Bar-Natan, online at <http://www.ma.huji.ac.il/~drorbn/Gallery/Symmetry/Tilings/S333/Kaleidoscope.html>.

Reflections can be used to trap light in an object. When a gem such as a diamond is cut into the shape of a polyhedron, it gives light an opportunity to reflect many times once it is captured inside. One of the reasons that a diamond is precious is its ability to bend light so that it stays inside the gem longer, thus making it sparkle.

Sound waves reflect in a theater to amplify music. Prior to electronic amplifiers, which increase the volume of microphones and electric guitars at rock concerts, special attention was paid to acoustical architecture in concert halls. Next time you watch a performance or a symphony in an indoor theater, notice the special plates built in or attached to the ceiling. They are angled and curved in order to reflect sound waves so that everyone in the theater can hear the performance. Without this special attention to reflecting sound waves, certain sections of the concert hall would not receive adequate sound, because the sound would either be absorbed by a surface, dissipate, or create destructive interference patterns. (See ***Inverse Square Function***.)

Reflections are also used in remote sensors to detect a signal. For example, there are several ways that you can change your television station using a remote control. One way is to aim the remote so that its ray will land directly on the sensor on the television set. Another way, however, is to aim the remote at a reflection of the sensor. Imagine that one of the walls in your home was a reflecting mirror, and determine the location of the television sensor behind the wall. If you aim the remote at the reflection of the sensor, the light beam will bounce off of the wall and land directly on the sensor. Many motion-based security systems operate in a similar fashion. An invisible beam reflects off of all walls in a room, creating multiple beams throughout that room. The alarm system is signaled if the beam at any point in the room is disturbed.

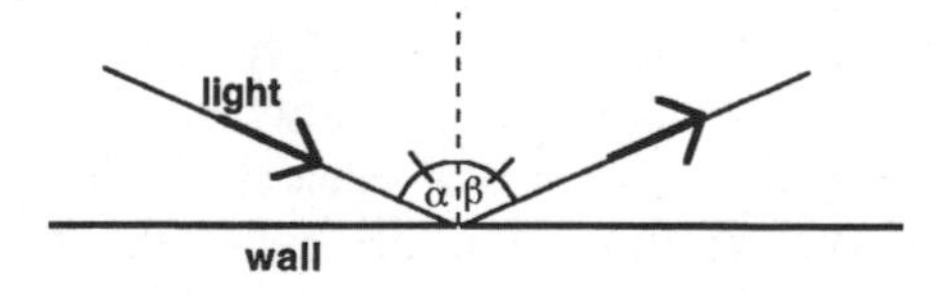

Light rays reflect from walls at congruent angles. The angle of incidence, α, has the same measure as the angle of reflection, β.

The angle of incidence, α, is the angle at which a beam of light touches a wall, and the angle of reflection, β, is the angle at which the beam leaves the wall. If the beam of light does not pass through the material, then the angle of incidence is equal to the angle of reflection. (See ***Angle*** for more explanation.) Knowing this theorem can help you become skilled at various games that use reflections, such as billiards and miniature golf. In both of these activities, the player is usually at an advantage if he or she can find ways to maneuver the ball by bouncing it off of a wall. In order to accurately place a ball on a target or in a hole, the player needs to aim the ball towards the reflection of the hole, similar to directing a remote control. Therefore an easier way to utilize the reflection is to predict the location on the wall where the angle of incidence will equal the angle of reflection.

Using reflections in games becomes more complicated, however, in situations where a ball needs to bounce off two walls. The same relationships regarding reflections exist in these circumstances, but the player will need to focus on reflections of reflections in order to utilize multiple walls in the attempt.

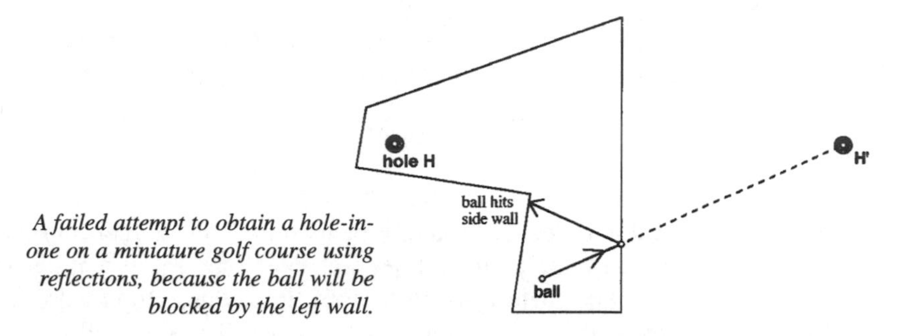

A failed attempt to obtain a hole-in-one on a miniature golf course using reflections, because the ball will be blocked by the left wall.

For example, suppose the player realizes that he or she cannot get a hole-in-one by hitting just one wall, as indicated in the above figure. Instead, the player imagines hitting two walls, the side wall first and then the back wall. In order to sink the shot, he or she will need to locate the reflection of the hole on the back wall, H', and then the location of the reflection of the reflection, H''. The player then aims towards the side wall at the double reflection of the hole, H'', and the ball should follow a path towards the first reflection by hitting the back wall, and then land in the hole, as shown below.

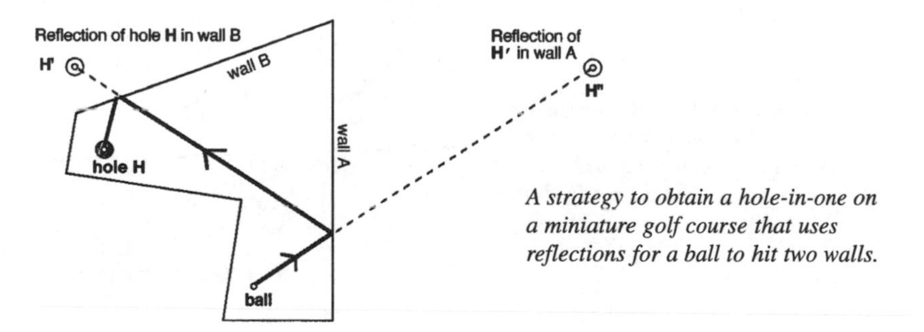

A strategy to obtain a hole-in-one on a miniature golf course that uses reflections for a ball to hit two walls.

The process can get even more complicated with more reflections, such as what takes place in games like racketball. In such a fast game, it may be difficult to predict where the ball will eventually land after it has been struck. However, a general knowledge of reflections can give a player a sense of what direction the ball will head once it hits the first wall.

The relationship between the angle of incidence and angle of reflection also informs product designers that full-length mirrors should only be one-half a person's height. In this type of mirror, the reflection of light from your eye level to your waistline will angle down towards your toes (see the following figure). This relationship is true, because the point of contact with your line of sight and the mirror is at the midpoint of your body, where the angle of incidence is congruent to the angle of reflection. That way, looking at the bottom of a mirror that is half your size will allow you to look directly at your feet.

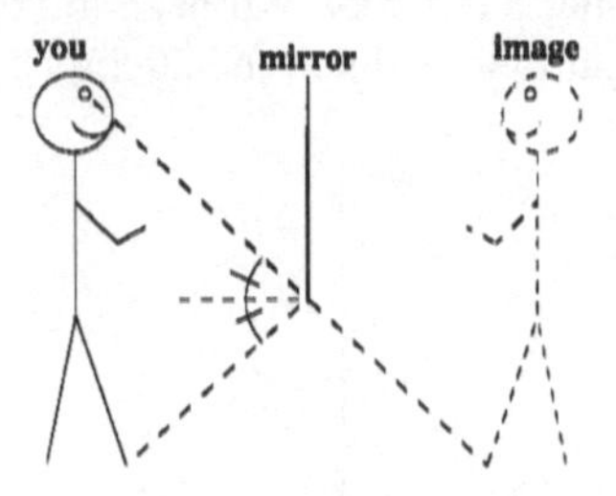

A mirror needs to be only one-half your height in order to see your entire body.

The concepts behind reflections can also be used to optimize fuel consumption in water travel. Suppose a cruise liner was departing a port and headed towards a series of remote islands. Along the way, it may need to refuel near a mainland to ensure that it can travel the entire distance. The ship will be most fuel efficient if it angles its navigation towards the shore to refuel, so that its angle of incidence is equal to its angle of reflection. Even though the ship will not use a reflection, moving along this path allows it to travel the smallest distance, as shown in the figure below. This path will be equivalent in distance to a direct route between the starting point and destination, because the ship will be directed towards the reflection of the destination. Since reflections preserve congruence, the ship will still be traveling along a line, which is the shortest path between two points.

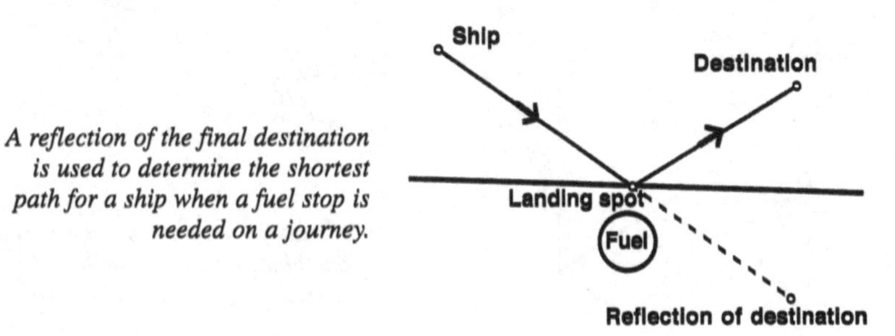

A reflection of the final destination is used to determine the shortest path for a ship when a fuel stop is needed on a journey.

Most molecules come in two forms, mirror images of each other. This would be merely a chemical curiosity were it not that the reflection images of molecules in medications can produce catastrophically different results. During the 1960s, the drug thalidomide was given to pregnant women to reduce nausea. One of the forms of the thalidomide molecule worked well for that task. Unfortunately, the other form of thalidomide, the mirror image of the good molecule, caused limb deformities in thousands of infants before its effects were recognized. The drug L-dopa counteracts symptoms of Parkinson's disease such as tremors and muscle rigidity. The mirror image of the L-dopa molecule, D-dopa, is toxic. The production of thalidomide and L-dopa produces both forms in equal amounts. A significant task for scientists was to determine how to remove the toxic form of the molecules from medications, leaving only the beneficial mirror images. The

2001 Nobel prize in chemistry was awarded to scientists who developed catalysts that would determine the twists in the molecules and either remove the malignant forms or change their orientation to the beneficial version.

online sources for further exploration

The billiards simulation
<http://serendip.brynmawr.edu/chaos/home.html>

Diamond design
<http://www.gemology.ru/cut/english/tolkow/_tolk1.htm>

Frieze patterns
<http://www.ucs.mun.ca/~mathed/Geometry/Transformations/frieze.html>

Mathematical art of M. C. Escher
<http://www.mathacademy.com/pr/minitext/escher/>

Measurement microphones
<http://www.josephson.com/tn6.txt>

Reflection of light
<http://micro.magnet.fsu.edu/primer/lightandcolor/reflection.html>

Reflectors
<http://nths.newtrier.k12.il.us/academics/math/Connections/reflection/REFLECT.htm>

Road coatings
<http://www.montefiore.ulg.ac.be/services/acous/eclair/reflecten.html>

Seismic reflection
<http://www.enviroscan.com/techapps/22.html>

Total internal reflection
<http://www.sciencejoywagon.com/physicszone/lesson/09waves/totint/>
<http://www.glenbrook.k12.il.us/gbssci/phys/Class/refrn/u14l3c.html>
<http://www.sciencejoywagon.com/physicszone/lesson/09waves/totint/internal.htm>

▲▼▲

ROTATIONS

Rotations allow the same object to reappear along a circular path. For example, in a tiling pattern, lizards can be rotated so that they tessellate, or fit neatly into each others' grooves, as shown in the following figure. Since there are three congruent lizards in a circle centered around the intersection of the lizard's knees, the angle of rotation must be one-third of the degree measure of a circle, which is 120 degrees.

Rotations are used in circular motion, such as the rotation of a wheel caused by movement in the axle of a car. The rotation of the wheels and the friction between the wheels and the road enable a car to move forward.

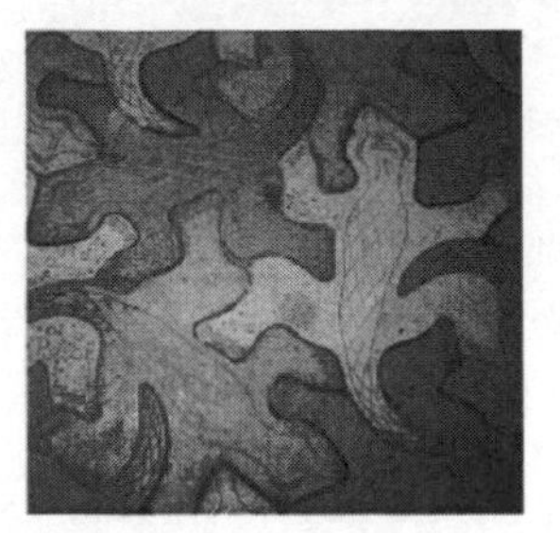

Tessellation of lizard tiles based on rotations for an outdoor patio.

Circular gears with wedges help support rotation in engines and machines. For example, a bicycle uses gears to change the amount of force needed to move the pedals. As the gear rotates, its teeth grab onto the chain and move it forward in order to spin the wheels on the bicycle. Gears with a smaller radius require less force, since the chains move a smaller distance. As the bicycle builds speed, the gears rotate more quickly, making it more difficult to pedal in lower gears. By shifting the chain to a higher gear with a greater radius when the bicycle increases speed, the pedals will slow down, since the chain has a greater distance to move, making it easier to maintain a higher speed. When the person on the bicycle slows down, the gears should be shifted down to a smaller radius so that pedaling becomes easier.

The gears of a bicycle rotate and latch onto the chain to help propel the bicycle forward when force is applied to the pedals.

Several amusement-park rides rotate to create a spinning effect. An object will feel like it is moving more quickly around a circle if it sits further away from its center of rotation. In this situation, the object has to travel a further distance around a circle than an object closer to the center, but also during the same time period. Rotational motion with ice skaters changes, because angular momentum is conserved. Angular momentum is determined by the product of the radius of the arm length and the skater's angular velocity. If his or her arm radius decreases so that the arms are closer to the body, his or her angular velocity will increase. As a result, skaters will spin faster when they move their arms closer to their bodies. Conversely, skaters can slow down their spinning motion by spreading their arms out from their bodies.

The earth rotates around an axis that passes through the two poles. The radius of the earth is 3,963 miles. Therefore every object at the equator is moving at a rate of approximately 1,037 miles per hour, because these objects travel $2\pi(3{,}963)$ miles in twenty-four hours. The angular rate of objects in circular motion is the circular distance divided by the time to travel that distance. If a person is standing away from the equator, then his or her angular rate is $\frac{2\pi \bullet 3963 \bullet \cos\theta}{24}$, where θ is the latitudinal angle of the city. For example, if a person is standing at 60°N latitude, then he or she will only be half as far from the earth's axis of rotation, because cos 60° is one-half. Then this person will only be moving half as fast around the earth. People actually do not feel like they are moving faster at different parts of the world because everything else is moving at the same rate. You feel differences in motion when something else is moving faster or slower than your motion.

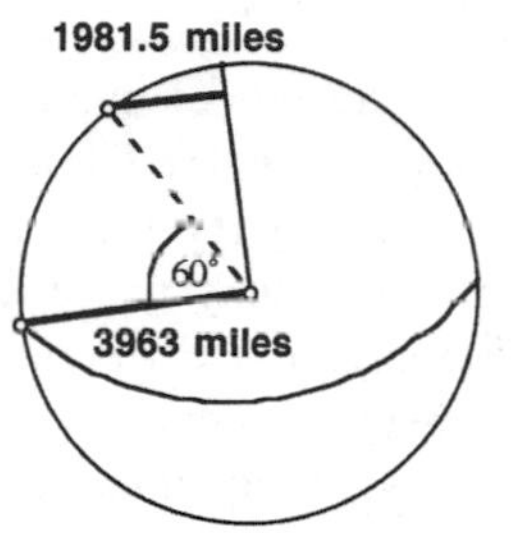

Relative distance away from the earth's axis of rotation based on latitudinal position. A person is half as far from the earth's axis of rotation when he or she is standing at 60° latitude because cos 60° = 1/2.

Rocket launches take advantage of the earth's rotational velocity. The launched aircraft takes off towards the east—the same direction as the rotation of the earth—giving it an extra boost once it is airborne. Also, launches in the United States are at Cape Canaveral, Florida, which is closer to the equator than most other cities in the country. Since it is farther from the earth's axis of rotation than many other U.S. cities, the earth's rotation will be more supportive at that location by giving it almost the best possible additional speed as it leaves the earth's atmosphere.

online sources for further exploration

Applications of rotations
<http://www.spacesciencegroup.nsula.edu/sotw/newlessons/application.asp?Theme=astronomy&PageName=rotation>

Bicycle gears
<http://www.exploratorium.edu/cycling/gears1.html>
<http://www.exploratorium.edu/cycling/gears3.html>

Image rotation
<http://www.ece.gatech.edu/research/pica/simpil/applications/rotation.html>

Mathematical art of M. C. Escher
<http://www.mathacademy.com/pr/minitext/escher/>

Relative motion—rotation and the motion of the moon
<http://www.joma.org/vol1-2/modules/macmatc4/moon.html>

Rotations on a sphere
<http://www.uwgb.edu/dutchs/mathalgo/sphere0.htm>

Satellite reception
<http://repairfaq.cis.upenn.edu/sam/icets/satellte.htm>

X-ray diffractometry
<http://www.optra.com/XRAYwebsite.htm>

▲▼▲

SEQUENCES

Sequences are sets of numbers that often share a recursive or explicit relationship. For example, the Fibonacci sequence in the form 1, 1, 2, 3, 5, 8, 13, 21, . . . is determined by the sum of every two previous consecutive integers in its sequence and has many real-world applications. (See ***Fibonacci Sequence*** for several examples.) A different pattern occurs in the terms in a geometric sequence, where consecutive terms have a constant ratio. A geometric sequence with an initial value equal to 4 and constant ratio of –0.5 would be 4, –2, 1, –0.5, 0.25, . . . Another type of sequence based on a constant difference between terms is called an *arithmetic sequence*. An arithmetic sequence with an initial value equal to 4 and a constant difference of –0.5 would be 4, 3.5, 3, 2.5, 2, . . . Sequences exist in applications that have discrete and predictable patterns, such as the value of an automobile, camera aperture, music notes, or predicting the timing of an eruption.

Automobile value is based on its original price, depreciation rate, and age. Since the depreciation is fairly constant for a particular model, a car's yearly prices can be determined using a geometric sequence. The constant ratio in this case is 0.80, since the car maintains 80 percent of its value after each year. A car selling for $20,000 new that depreciates 20 percent each year will be worth $16,000 the next year, and $12,800 the year after that. These values can be determined by multiplying each successive term by 0.80, or using the explicit formula for a geometric sequence, $g_n = g_1 r^{n-1}$, where g_n is the value of the car after the *n*th year, g_1 is the initial value of the car during the first year, and r is the constant ratio. In this case, the explicit equation for the sequence is $g_n = 20,000(0.80)^{n-1}$. The table on the next page represents a sample blue-book listing of the value of a vehicle for different years based on this equation. Notice that the car loses its greatest amount of value during the first year, since a percentage of the total value is reduced from the original price.

Standard f-stops on cameras permit the photographer to select how much light passes through the lens. The sequence is 1, 1.4, 2, 2.8, 4, 5.6, 8, 11, 16, 22, 32. Each of the f-stop numbers on a standard lens represents half the light of the number before it. The consecutive f-stops are in geometric sequence with the common ratio $\sqrt{2}$.

nth year	*value ($)*
1	20,000.00
2	16,000.00
3	12,800.00
4	10,240.00
5	8192.00
6	6553.60
7	5242.88
8	4194.30
9	3355.44
10	2684.35

The estimated values of an automobile with a new price of $20,000 and depreciating 20 percent each year.

The twelve tones in an octave form a geometric sequence so that the end of an octave has a frequency twice that of its low tone. High C (512 cps) is twice middle C (256 cps). The multiplication of frequencies is a constant ratio across the octave, so each multiplication must be the twelfth root of 2, or about 1.059. So if A is 440 cps, the next key, B-flat, will be $440 \cdot 1.059 \approx 466$ cps. (See ***Ratio***.)

Old Faithful is a popular attraction at Yellowstone National Park, because the geyser produces long eruptions that are fairly predictable. When tourists visit Old Faithful, they will see a sign that indicates an estimated time that the geyser will next erupt. No one controls the geyser like an amusement park ride. Instead, its patterns over time have caused park rangers to develop predictable eruption times using an arithmetic sequence. The time between eruptions is based on the length of the previous eruption. If an eruption lasts one minute, then the next eruption will occur in approximately forty-six minutes (plus or minus ten minutes). If an eruption lasts two minutes, then the next eruption will occur in approximately fifty-eight minutes. This pattern continues based on a constant difference of

Old Faithful is a popular attraction at Yellowstone National Park, because it is a very large geyser and quite predictable. Source: National Park Service photograph.

twelve minutes, forming an arithmetic sequence of 46, 58, 70, 82, 94, . . . An eruption of n minutes will indicate that the next eruption, a_n, will occur in $a_n = a_1 + (n - 1)d$ minutes, where a_1 is the length after a one-minute eruption, and d is the constant difference of waiting time among eruptions that are a one-minute difference in time. In this particular situation, the next eruption will occur in $a_n = 46 + (n - 1)12$ minutes, if the previous eruption was n minutes long.

Harmonious chords produce another type of sequence. If you depress piano keys for middle C, middle G, high C, and high E, then play low C, you will hear the four other tones. If the string for low C is one meter long, then a string a half meter long will sound a middle C, a one-third meter string would give middle G, high C would be one-fourth meter, and high E would be one-fifth meter. The list of overtones is the sequence $1, \frac{1}{2}, \frac{1}{3}, \frac{1}{4}, \frac{1}{5}, \ldots$ which is called a *harmonic sequence*. Any sequence that is formed from reciprocals of terms of an arithmetic sequence is called *harmonic*. Harmonic sequences are important in the study of magnetism, electricity, and the design of electric motors. Cosmologists studying the origins of the universe look for harmonic patterns in microwave traces received from space.

online sources for further exploration

A demonstration that the harmonic series doesn't converge
<http://www.mathematik.com/Harmonic/>

Artificial intelligence
<http://www.cs.wustl.edu/area-ai.html>

Biological sequence alignment
<http://www.ics.uci.edu/~eppstein/gina/align.html>

DNA sequence database
<http://www.ncbi.nlm.nih.gov/collab/>

Intensity, exposure, and time in photography
<http://www.arch.virginia.edu/arch569/content/lectures/lec-03/>

Iteration and recursion activities
<http://www.colorado.edu/education/DMP/activities/iteration_recursion/>

Linear models
<http://www.math.montana.edu/frankw/ccp/modeling/discrete/linear/learn.htm>

Musical scales
<http://www.tromba.demon.co.uk/scales.html>
<http://www.midicode.com/tunings/greek.shtml>

Predicting Old Faithful
<http://www.jason.org/expeditions/jason8/yellowstone/oldfait1.html>

Used car prices
<http://www.edmunds.com/used/>
<http://www.kbb.com/kb/ki.dll/kw.kc.bz?kbb&&688&zip_ucr;1409&>

SERIES

Many applications that are based on the sum of predictable discrete patterns can be examined with *series*. For example, a doctor may prescribe an amount of medication to take each day, because he or she knows that the patient's bloodstream will be able to maintain a certain level of the medication over time. Prescriptions are based on a mathematical series, because the total amount of drug accumulates in the bloodstream each day. In other words, the sum of the remaining amounts of the drug in the bloodstream is added to a new amount everyday. One way to determine the total amount of a drug that will eventually end up in the bloodstream is to take the initial amount and add the amount that remains from yesterday, from two days ago, three days ago, and so on. If the amount of drug that remains in the bloodstream is a predictable pattern each day, then an equation can be used to compare dosages and accumulating amounts in the bloodstream.

Some illnesses, such as high blood pressure or thyroid deficiency, can be treated with regular medication. Suppose a doctor knows that 200 mg of a drug is the amount of medication needed to maintain the patient's health. Because most drugs circulate in the bloodstream, amounts of the drug are removed as the blood is cleaned by the kidneys. Suppose that the kidneys remove 40 percent of the drug each day. That leaves the drug effectiveness at 60 percent of what it was twenty-four hours earlier. Therefore the doctor has the patient take a pill each day. Surprisingly, a 200 mg pill each day is far too large a dose to maintain a 200 mg level in the bloodstream. If the doctor prescribes 200 mg each day, the patient will have 200 mg in the bloodstream on the first day. At the end of one day, only 120 mg will remain, but another 200 mg will be added, making the total amount 320 mg. This overdose can potentially be very harmful for the patient, so the doctor needs to determine an ideal dosage that will allow only 200 mg to remain in the bloodstream at any given time.

A pharmacist can model this situation by using a spreadsheet or table of values, making sure that the amount in the bloodstream at the end of the day is 60 percent the amount at the beginning of the day, and then adding that value to the amount at the beginning of the next day. The following table illustrates how much of the drug would remain in the bloodstream during the first twenty days if 200 mg were taken each day. Notice that eventually the amount of drug in the bloodstream will level off near 500 mg after about ten days.

days	*start of day*	*end of day*
1	200.000	120.000
2	320.000	192.000
3	392.000	235.200
4	435.200	261.120
5	461.120	276.672
6	476.672	286.003

days	*start of day*	*end of day*
7	486.003	291.602
8	491.602	294.961
9	494.961	296.977
10	496.977	298.186
11	498.186	298.912
12	498.912	299.347
13	499.347	299.608
14	499.608	299.765
15	499.765	299.859
16	499.859	299.915
17	499.915	299.949
18	499.949	299.970
19	499.970	299.982
20	499.982	299.989

The amount of drug (in milligrams) in a person's bloodstream when 60 percent remains from the previous day and an additional 200 mg are added each day.

A pharmacist can modify this initial amount on the first day and observe changes in the limit of this sum to determine that 80 mg is an appropriate daily dosage to maintain 200 mg in the bloodstream over time, as shown below.

days	*start of day*	*end of day*
1	80.000	48.000
2	128.000	76.800
3	156.800	94.080
4	174.080	104.448
5	184.448	110.669
6	190.669	114.402
7	194.401	116.641
8	196.641	117.984
9	197.984	118.791
10	198.791	119.274
11	199.274	119.565
12	199.565	119.739
13	199.739	119.843

days	*start of day*	*end of day*
14	199.843	119.906
15	199.906	119.944
16	199.944	119.966
17	199.966	119.980
18	199.980	119.988
19	199.988	119.993
20	199.993	119.996

The amount of drug (in milligrams) in a person's bloodstream when 60 percent remains from the previous day and an additional 80 mg are added each day.

This situation is an example of a geometric series, since the amount remaining in the bloodstream is affected by a constant ratio of 60 percent. The sum can be rewritten as

days since last dosage

	1	*2*	*3*	*4*	
80 +	$80(0.60)^1$ +	$80(0.60)^2$ +	$80(0.60)^3$ +	$80(0.60)^4$	+ . . .

The sum, s, can be determined by the equation $s = \frac{g_1(1-r^n)}{1-r}$, where g_1 is the initial dosage, r is the constant ratio, and n is the number of days the dosage is taken. Since the number of days that the drug is taken is unknown, pharmacists need to examine situations in which the drug is taken indefinitely. Therefore the sum of an infinite geometric series is $s = \frac{g_1}{1-r}$ because $\lim_{n \to \infty} \frac{g_1(1-r^n)}{1-r} = \frac{g_1}{1-r}$ when $|r| < 1$. In this case, the desired sum, s, is 200 mg, r is 60 percent or 0.60, and g_1 is unknown. Substituting the values into the equation, you will get $200 = \frac{g_1}{1-0.60}$, and a solution of $g_1 = 80$ mg. Thus the doctor needs to make prescriptions of 80 mg each day in order to maintain the desired dosage of 200 mg.

Geometric series are also used to predict the amount of lumber that can be cut down each year in a forest to ensure that the number of trees remain at a stable level. Each year, forest rangers plant seeds for new trees to account for those chopped down and lost to forest fires. Suppose the ranger wants to know what proportion of trees they can afford to lose or remove each year if they plant 500 new trees and want to consistently maintain 80,000 trees in the forest. After substituting s and g_1 in the formula $s = \frac{g_1}{1-r}$, the unknown value for r is 0.00625. This means that the forest ranger wants to maintain 99.375 percent of the trees each year. However, an interesting phenomenon is to notice that the forest can recover from a disaster such as a fire in a reasonably short period of time. Suppose a fire destroys 35 percent of the trees in the forest, leaving 52,000 trees. If 500 new trees are planted each year, and 0.625 percent of the total number of trees is used for wood each year, then the forest will reach its ideal level of 80,000 trees in about seven years, since $\frac{500(1-0.9375^7)}{1-0.9375} > 28,000$, the number of trees lost in the fire.

Arithmetic series are used when consecutive values have a constant difference. The sum of the first n terms of the series, s_n, is determined by the equation $s_n = (a_1 + a_n)(\frac{n}{2})$ or $s_n = (2a_1 + (n-1)d)(\frac{n}{2})$. For example, the sum of the first one hundred positive integers is generated by the series $1 + 2 + 3 + \ldots + 100$. This is an arithmetic series with $n = 100$ terms, $a_1 = 1$, and $a_{100} = 100$. Therefore the sum of this series, $s_{100} = (1 + 100)(\frac{100}{2}) = 5,050$. Drilling and mining operations use arithmetic series to determine the total distance their machines will need to drill when excavating rock from the earth. Suppose a construction team is hired to dig a hole that has a cross-sectional area of 10 square meters and will be 50 meters deep. Suppose that the drilling machine moves 2 feet downward when digging in the earth's surface, and then stops to allow workers to remove the 20 cubic meters of loose dirt. Therefore the first drilling attempt will be 2 meters deep, the second drilling attempt will be 4 meters deep, the third drilling attempt will be 6 meters deep, and so on for a total of 25 trips. In this situation, the drilling machine and the dirt from the ground will need to be moved a total of 650 meters in order to dig a 50 meter hole, since $(2 \bullet 2 + (25-1)2)(\frac{25}{2}) = 650$. The mining or construction company can then use this information to determine its fees based on the total distance it will need to move dirt out of the hole.

online sources for further exploration

Buying on credit
<http://www.nap.edu/html/hs_math/bc.html>

Drug dosage
<http://www.nap.edu/html/hs_math/drd.html>
<http://www.chch.school.nz/cma/IdeasTeach/hypnotic.htm>
<http://barzilai.org/cr/med-dosage.html>

Geometric series applications
<http://www.math.montana.edu/frankw/ccp/calculus/series/geometric/learn.htm>

Loan or investment formulas
<http://oakroadsystems.com/math/loan.htm>

SIMILARITY

Two figures are *similar* if they have the same shape, but not necessarily the same size. More specifically, all of the corresponding sides between two similar shapes are proportional and all of the corresponding angles are congruent. For example, most rectangular television screens are similar, since they have a 4-to-3 aspect ratio. That means that conventional television screens are produced so that the length is 4/3 times the width. The diagonal length of the television screen is often the reported number in advertisements. Using the 4-to-3 aspect ratio, a television screen that has a 25 inch diagonal will have dimensions of 16 inches by 9 inches, and a television screen with a 40 inch diagonal will have dimensions of 32 inches by 24 inches. Notice that the diagonal-to-length ratio is 5 to 4, and the diagonal to width ratio is 5 to 3, causing the width, length, and diagonals of every standard television set to be a multiple of the {3,4,5} Pythagorean triple.

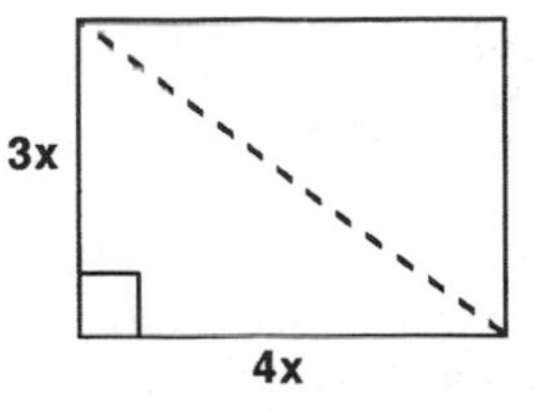

Dimensions of a standard television screen with a 4-to-3 aspect ratio.

In 1889, engineers in Thomas Edison's laboratory established that the 4:3 ratio was the best one for movie screens. It is now being challenged by the 16:9 ratio for high-definition TV sets (HDTV) that use a wider screen than the traditional one to mimic the wide screens in theaters.

Book covers are examples of two objects that are often not similar. Even though two books may have rectangular covers with congruent angles, they are only similar if their side lengths are proportional. For example, a book cover with dimensions of 6 inches by 3.5 inches is not similar to a book cover with dimensions of 7 inches by 4.5 inches. The corresponding ratios of 7/6 and 4.5/3.5, or 9/7, are not equal.

Similarity is used for many real-world purposes. The film on a movie reel is projected onto a big screen so that the images appear larger, but in the same proportion. If the screen images were not similar to the slides on the reel, the images would appear distorted, being either too fat or too long (see ***Proportions*** for a more detailed explanation). An overhead projector serves the same purpose, allowing images such as a teacher's handwriting to appear larger on a screen so that it is easier to read. A telescope and microscope also change the size of images, making them easier to see while preserving the shape of the original object. The development of pictures from a camera also uses similarity principles. As negatives are processed onto photo paper, they expand uniformly in size. If a picture needs to be enlarged into a poster, then the ratio of the corresponding sides between the negative and the poster need to be identical. This means that if the different sizes of photo paper are not similar, then some cropping will occur.

Similarity can be used to approximate lengths and distances. For example, on a sunny day you can use similarity to determine the height of a tall object such as a flagpole by using just a tape measure. If you measure your height, your distance from the flagpole, and the length of your shadow, then you will be able to set up a proportion to find the height of the flagpole. For instance, suppose you are standing 5 meters away from the flagpole, you are 1.65 meters tall, and you measure your shadow to be 1.38 meters long (see the figure below). Similar triangles can be used to show that your height corresponds to the flagpole height, and your shadow length corresponds to the flagpole's shadow length.

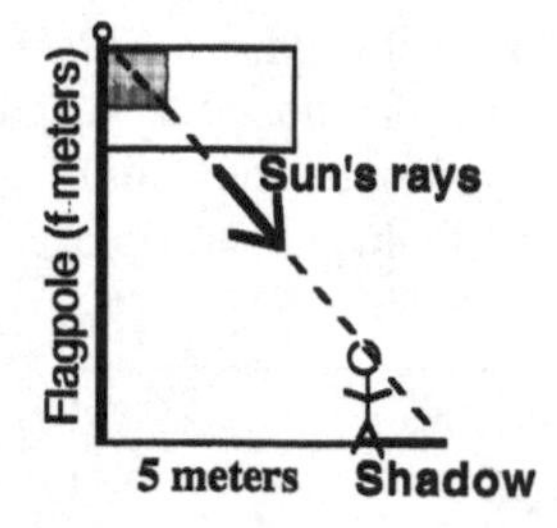

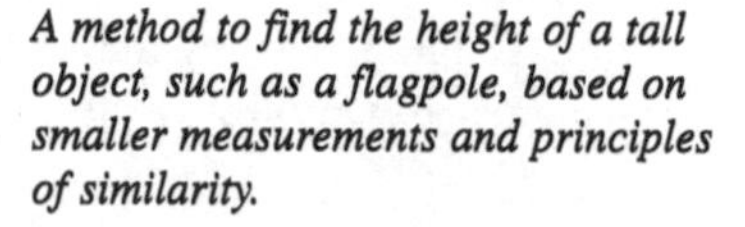
A method to find the height of a tall object, such as a flagpole, based on smaller measurements and principles of similarity.

In this case, the proportion $\frac{f}{5+1.38} = \frac{1.64}{1.38}$ can be used to find the height of the flagpole, f, which equals approximately 7.63 meters (close to 25 feet).

Architects and designers use similarity to create and visualize new buildings. A miniature two- or three-dimensional model that is a replica of a future building is often put together during a design phase. It is easier and less expensive to make changes to a miniature replica of an object than to the object itself, so careful attention to size and detail is important in model-making. Once the ideas behind the design of the house are negotiated, the floor plans are passed on to the builders to replicate the model on a larger scale. Since the actual floor space of the house is similar to the paper mock-up of the floor plan, the corresponding dimensions between the real structure and the model are proportional. However, the area comparing the house's floor space to the floor-plan area is proportional to the square of the ratio of the dimensions. For example, if the house is 50 times larger than the floor plan, then the area of the house is 2,500 (which is 50^2) larger than the floor plan. This area proportion of similar figures is squared, because area is a measurement of two dimensions. For example, suppose two similar squares have respective lengths of 2 and 100 cm. The area of the squares would be 4 cm^2 and 10,000 cm^2, respectively. Even though the ratio of their lengths is 100/2 or 50, the ratio of their areas is 10,000/4 or 2,500, which is the same as 50^2. Carpenters can use this information to determine the amount of wood and carpeting needed for the floors if they are not given the actual dimensions of the house.

Similarity can also be used to predict the mass of unusually large or even extinct animals, such as dinosaurs. A scale model of a dinosaur can be used to predict the actual volume of it, assuming that the ratio comparing the actual length to the model length is available. Suppose that an accurately scaled model of a tyrannosaurus with a length of 0.3 meters is used to determine its mass.

Since an actual tyrannosaurus was about 15 meters long, the ratio of the actual dinosaur to the model is 50 to 1, because 15/0.3 = 50. Use the density ratio of $\frac{\text{mass}}{\text{volume}}$ to determine the mass of the tyrannosaurus. Most animals and reptiles have a density near $0.95 = \frac{m}{v}$, so the mass of the tyrannosaurus can be calculated once the volume is found. The volume of the actual tyrannosaurus can be calculated by using the cube of the ratio of the lengths of the actual dinosaur to the model. The cube of the ratio is used, because volume is a measure of three dimensions. Therefore the volume of the actual Tyrannosaurus will be 50^3, or 125,000 times the volume of the dinosaur model.

You can measure the volume of an irregular object, such as a dinosaur model, by submersing it in a bucket of water. Place a bucket of water filled to the brim (and larger than the dinosaur model) inside a larger empty bucket. Drop the dinosaur model into the bucket of water, and the excess water will spill over the sides into the empty bucket. Pour the excess water into a graduated cylinder, which is a tool to measure the volume of water. This volume should be the same as the volume of the dinosaur model, because the model replaced the same amount of space in the bucket as the excess water. Suppose that the volume of the model is 61 milliliters. This means that the volume of the actual tyrannosaurus was about 125,000 times 61, or 7,625,000 milliliters, or 7,625 liters. Since density equals mass divided by volume, the equation $0.95 = \frac{m}{7{,}625}$ can be used to predict the mass, m, of the tyrannosaurus. Note that the units of density are kilograms per liter, so volume units are in liters and calculated mass units are in kilograms. The solution to the equation predicts the tyrannosaurus's mass to equal approximately 7,243 kilograms, which is about 16,000 pounds. That is the same as 100 people that have an average mass of 160 pounds. Most football coaches would like to recruit a tyrannosaurus for their teams!

Similarity is sometimes not used in models, which as a result can cause misconceptions about length and size. Most models of the solar system are inaccurately proportioned so that they can be easily stored, carried, and viewed within a reasonable amount of space. If a teacher wants to illustrate planetary motion on a solar-system model, he or she needs to be able to move the planets around fairly easily, and students need to see all of them. Realistically, however, this type of model is inaccurate, because the planet sizes vary tremendously and are spread apart by vastly different distances. For example, if an accurate scale model of the planets in the solar system were used in a classroom with the sun at the center of the room, then the first four planets would be within 227 cm of the center, and the remaining planets would be stretched out to almost 6 meters away! The large variability in distances among the planets would make it difficult to build a movable model that illustrates rotation around the sun. Furthermore, the volumes of the planets vary considerably. Large planets, like Jupiter and Saturn, have diameters that are about ten times larger than the earth. If the planets were built to scale, these giant planets would have to be a thousand times larger than the earth, because the ratio of volumes between similar figures is equal to the ratio of the cubes of their lengths. For visualization and instructional purposes, this would be difficult to create in a hand-held model. It unfortunately provides misconceptions about the relative sizes and distances among planets in our solar system.

online sources for further exploration

Map making
<http://www.sonoma.edu/GIC/Geographica/MapInterp/Scale.html>
<http://www.epa.gov/ceisweb1/ceishome/atlas/learngeog/mapping.htm>

Nuclear medicine
<http://www.math.bcit.ca/examples/ary_11_1/ary_11_1.htm>

Scale model of a pyramid
<http://www.pbs.org/wgbh/nova/pyramid/geometry/model.html>

Scale models
<http://www.faa.gov/education/resource/f16draw.htm>
<http://www.pbs.org/wgbh/nova/pyramid/geometry/model.html>
<http://www.americanmodels.com/sscale.html>

Screen ratios
<http://www.premierstudios.com/ratio.html>
<http://www.pbs.org/opb/crashcourse/aspect_ratio/>

Understanding scale speed in model airplanes
<http://www.astroflight.com/scalespeed.html>

SLOPE. See **LINEAR FUNCTIONS; RATES**

▲▼▲

SQUARE ROOTS

A *square root* is the inverse of a squared number. The square root of 49, written as $\sqrt{49}$ or $49^{1/2}$, is equal to 7, because 7^2 equals 49. Many real-world relationships involve square roots. For example, the height of liquid wax in a candle is directly proportional to the square root of the amount of time a candle has been burning. This information is useful in the design of candles, because the presence of liquid will slow down the burning of the wick. Hence, fatter candles do not need very long wicks, because they will likely form a pool of liquid as they burn.

Pilots of airplanes and hot-air balloons use square roots to estimate viewing distances. The viewing distance in kilometers, d, from an airplane on a clear day, depending on its altitude in meters, a, can be estimated by the equation $v = 3.56\sqrt{a}$. The viewing distance from an airplane to the horizon is perpendicular to the radius of the earth, forming a right triangle between the airplane, horizon, and center of the earth (see the figure below). The Pythagorean theorem can be used to compare the distances, $v^2 + 6380^2 = (6380 + \frac{a}{1000})^2$, using the fact that the radius of the earth is 6,380 km. The square-root version of the equation is approximately equal to this format, since commercial airplanes do not fly much higher than 10,000 meters.

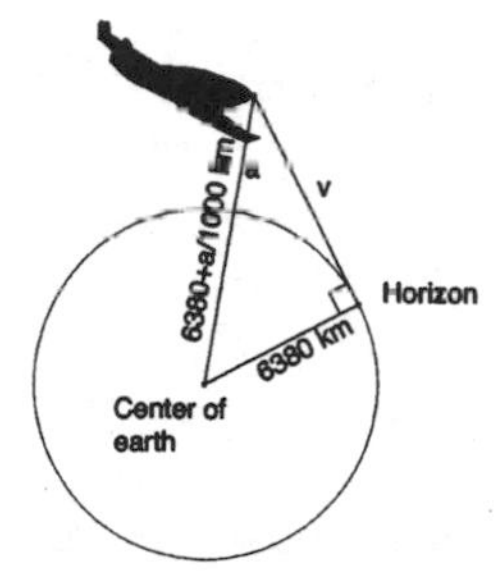

The Pythagorean theorem can be used to determine the maximum viewing distance, v, a pilot has in an airplane compared with its altitude, a. Note: this drawing is not to scale.

Police investigators use square roots at the scenes of auto accidents. They can estimate the speed of a car by the length of the tire skids and the conditions of the road. The speed of a car in miles per hour, s, that skidded d feet is $s = \sqrt{30fd}$. The variable f describes the coefficient of friction of the road. On dry concrete, this value is about 0.8, and in wet conditions, f is about 0.4. Measuring the length of the skids will help determine if the offender's speed was a factor that contributed to the accident.

The period of a pendulum, or the time it takes to move back and forth, can be determined by the equation $t = 2\pi\sqrt{\frac{l}{g}}$, where t is the time in seconds, l is its length in meters, and g is the acceleration due to gravity (9.8 meters/second^2). This equation is actually a combination of a couple of equations, $g = lw^2$ and $w = \frac{2\pi}{t}$, that relate to circular motion and the pendulum's length, period, and angular velocity w. Notice that the mass of the object at the end of pendulum is not included in the equation, because all objects will fall at the same rate, regardless of their mass. The pendulum equation is useful for clockmakers, because a grandfather clock is designed so that its pendulum arm takes one second to swing in one direction, or two seconds to swing back and forth. If $t = 2$ is substituted into the equation, then the pendulum arm length l will be approximately 1 meter long.

Using square roots can help a person become a better consumer of art. The best view of a picture is when the angle, α, from the bottom of the picture to the top is greatest, as shown in the following figure. An ideal distance, d, to stand from a painting is based on how much higher the bottom of the painting is from a person's eye level, b, and how much higher the top of the painting is from a person's eye level, t, according to the equation $d = \sqrt{bt}$.

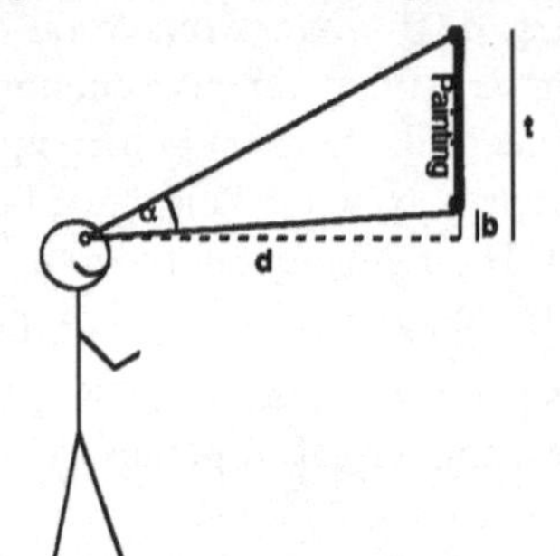

The optimum viewing angle of a painting, α, is greatest when the distance a person stands from the painting, d, is equal to the square root of the product of the distances from the edges of the painting to the eye-level height ($d = \sqrt{bt}$).

If the average human is about 67 inches tall, and a large painting is 60 inches tall, and the bottom is placed about 70 inches from the ground, then $b = 70 - 67 = 3$, and $t = 3 + 60 = 63$. Therefore a reasonable place to draw a viewing line would be about 14 inches away from the wall based on evaluating $d = \sqrt{3 \cdot 63} \approx 13.7$ inches. This formula can be applied to similar venues, such as helping you find the best seat in a movie theater.

online sources for further exploration

Best angle of view
<http://jwilson.coe.uga.edu/emt725/Angle.View/Angle.view.html>

Bouncing ball
<http://www.sosmath.com/calculus/geoser/bounce/bounce.html>

Calculating angles in a pyramid
<http://www.math.toronto.edu/mathnet/questionCorner/miter.html>

Distance between two ships
<http://www.nadn.navy.mil/MathDept/cdp/relatedrates/rates.html>

How to compute baseball standings
<http://www.math.toronto.edu/mathnet/questionCorner/baseball.html>

Latitude and longitude
<http://daniel.calpoly.edu/~dfrc/Robin/Latitude/distance.html>

The square root law of inventory
<http://logistics.about.com/industry/logistics/library/blsqrootlaw.htm>

▲▼▲

STANDARD DEVIATION

The *standard deviation* is a number that indicates the variability in a set of data. It is a measure of the dispersion of data in a sample or population. Standard deviations are used in quality control in business and industry and in the computation of standard test scores (such as the SAT and ACT). The concept of standard deviation provides the basis for widely used statistical techniques.

The start of the computation of standard deviation is the deviation about the mean, the difference of the actual score and mean score. If a college-placement test has a national mean of 512, and a student has a score of 650, the deviation is 138. Deviations are negative when the score is below the mean.

Even though each deviation tells something about the spread of data, the sum of deviations is always zero, which gives no overall information about the spread of the data. To make sure negative deviations do not cancel with positive, statisticians choose to square each deviation. Then they average the squared deviations to produce a number that indicates how the data is spread out around the mean. The average squared deviation is called the *variance*. The square root of the variance is the standard deviation. There are two formulas for standard deviation. One form assumes that the data set is the entire population of cases: $\sigma = \sqrt{\frac{\Sigma(X-\mu)^2}{N}}$, where μ is the mean of the data, and N is the number of pieces of data. If the numbers could be considered a *sample* from the population, then the mean and standard deviations would represent estimates of the entire season's scores. The standard deviation has a different symbol in this case, and a slightly different formula: $s = \sqrt{\frac{\Sigma(X-\overline{X})^2}{n-1}}$, where $\overline{X}$ is the mean of the sample, and n is the sample size.

The standard deviation is used to compute standardized scores for the comparison of data from different sets and measures. A standardized score is computed as $z = \frac{X-\mu}{\sigma}$, or the deviation divided by the standard deviation. As a ratio, it has no units. The standardized score can compare different measures of the same person. Suppose a student had a score of 540 on the SAT-Math and 24 on the ACT Mathematics. On which did he or she do better? The national mean for SAT-Math is 514, with a standard deviation of 113. So $z_{\text{SATM}} = \frac{540-514}{113} \approx 0.23$. The national mean for ACT Mathematics is 20.7, with a standard deviation of 5.0. So $z_{\text{ACTM}} = \frac{24-20.7}{5.0} \approx 0.66$. Therefore she did relatively better on the ACT Mathematics, because she had a greater standardized z score.

Z scores have been used to compare baseball players from different eras. Does Ty Cobb's batting average of .420 in 1922 represent better batting than George Brett's .390 in 1980? It has been argued that it is difficult for a player today to hit over .400, because the general quality of players is much higher than it was in the early days of professional baseball. If you use the standard scores based on means and standard deviations of baseball players in their respective eras, Cobb has a z score of about 4.15 and Brett, 4.07. The two stars were equally outstanding in performance during their respective eras.

SAT and ACT scores are normally distributed, which means that a frequency chart or histogram will appear to be bell-shaped. In this type of distribution, there are some handy "rules of thumb" that use standard deviation to describe the spread of data. In a normally distributed set of data, about 68 percent of it is contained within one standard deviation of the mean (as shown in the figure below), 95 percent within two standard deviations, and 99.7 percent within three standard deviations.

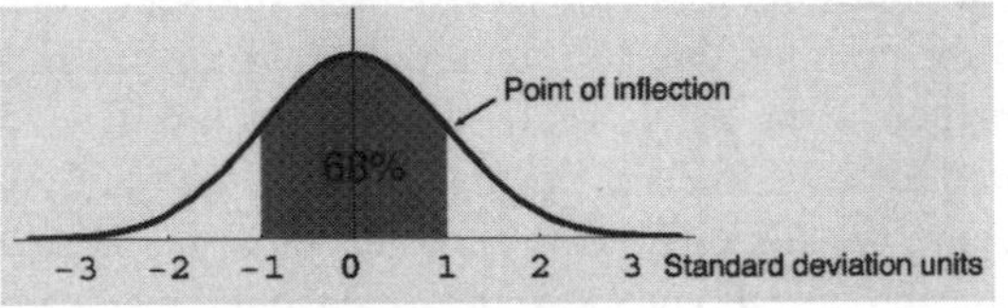

About 68 percent of the area under a normal distribution curve is between -1 and 1 standard deviations (z scores).

The normal curve has two points of inflection where the curve changes from concave-downward, to concave-upward. These are located at ±1 standard deviation units. The point of inflection at +1 standard deviation is shown in the figure.

The rules of thumb for a normal distribution stop at ±3 standard deviations from the mean, because almost all of the data is trapped by those limits. That is not enough for the management goal of "six sigma" quality adopted by many American businesses. In such cases, the goal is to have fewer than 3.4 defects per million products. The six sigma, or 6σ, is chosen because 99.99966 percent of the cases in a normal distribution fall within six standard deviations of the mean. If that proportion represents defect-free products, then the remainder, 0.00034 percent, represent defects. Such high-quality control standards at six sigma will likely reduce the number of defects in a product, but at a high cost when an inspection fails. Reducing the standards to 99.7 percent defect-free products will likely save the company money in the long run, unless the company is dealing with personal health and safety issues. Physicists use a five-sigma criterion in determining whether a subatomic particle has been revealed. They think that only a five-sigma result, indicating a 99.99995 percent chance that the result can be reproduced, is trustworthy and can survive the test of time.

The rules of thumb are often used by manufacturers to design clothing and furniture that will sell to the broadest audience. For example, an automobile manufacturer developing an automobile for potential female customers might design the driver's seat to fit the heights of most women. To make the greatest profit, the seat must be as standard as possible. The heights of American women are normally distributed with a mean of 64 inches, with a standard deviation of 2.5 inches. If the manufacturer has its designers work on a seat that will be comfortable for women from 59 to 69 inches tall (two standard deviations above and below the mean), then the rule of thumb says that the seat would be appropriate for 95 percent of the women.

In medical quality-control testing it is difficult to evaluate the effectiveness of a medical instrument, because many medical measurements such as blood pressure, glucose content in urine, and cholesterol in blood can have different distributions based on sex or age. Some electronic sensors have the statistics for different population groups in memory. When a reading for a particular type of patient is more than two standard deviations from the mean for his or her group, the instrument will sound a tone, alerting nurse or doctor to the critical value. A dynamic instrument that accounts for patient's variables establishes a more precise diagnosis of medical problems.

When the standard deviation is computed from statistics on many samples, such as a standard deviation of ACT composite school averages for many schools, the standard deviation is called a *standard error*. Survey statistics in newspapers are often reported as a range of values, such as in "our survey of 250 randomly selected adults showed that 62 percent of the residents oppose the new highway. The margin of error was 6 percent." In most cases, the margin of error for a reported statistic is two standard errors. The report of the survey results would be "62% ± 2•Standard Error." This gives a range of values that is likely (95 percent certain) to trap the percentage that would have been obtained had the entire population been surveyed. So the newspaper would be saying, "If the entire population of residents had been surveyed, there is a 95 percent chance that the true proportion is between 56 percent and 68 percent." In the weeks prior to national and state elections, you will read about polls that indicate which candidate is ahead in the race, and whether the candidate has a clear lead. If candidates are separated by two standard errors, the newspaper would project a winner. The sampling of voters as they leave polling booths is a method that television networks have used to make predictions of winners on their news programming shortly after the polls close. However, as the networks found out in the November 2000 presidential election, it is necessary that samples be carefully designed to be representative of the population. Had the networks followed the cautious recommendations of statisticians, they would not have had to make their embarrassing switches of victory reports from George Bush to Al Gore based on the controversial voting reports from the state of Florida.

Statistics computed on samples establish the close connection between standard deviation and the normal curve. Although the numbers in an entire population might not follow a normal distribution, the *central limit theorem* states that *means* of samples from the population will be normally distributed. Further, the standard deviation of the sample means (standard error of the mean) is the standard deviation of the population divided by the square root of the sample size. The central limit theorem is the foundation for *inferential* statistics, the branch of statistics that is used to determine whether a new drug is better than older treatments, whether consumers really like the flavor of a new, improved toothpaste, when an assembly line is producing too many defects, whether students in a school are not doing well on a state test, and when a stock price is stabilizing. Pollsters use the central limit theorem to determine how large their samples must be to reach a desired level of accuracy.

online sources for further exploration

Baseball
<http://www.stat.ncsu.edu/~st350_info/reiland/350hw3.htm>

Biomedical electronics
<http://www.math.bcit.ca/examples/ary_1_8/ary_1_8.htm>

Bioretention applications
<http://www.epa.gov/nps/bioretention.pdf>

Election polls
<http://www.pollingreport.com/election.htm>

Estimating trees
<http://www.math.bcit.ca/examples/ary_15_8/ary_15_8.htm>

Food technology
<http://www.math.bcit.ca/examples/ary_2_8/ary_2_8.htm>

Gallup polls
<http://www.gallup.com/>

Indiana custom rates
<http://www.agecon.purdue.edu/extensio/pubs/custom_rates.htm>

Mining
<http://www.math.bcit.ca/examples/ary_10_8/ary_10_8.htm>

Petroleum technology
<http://www.math.bcit.ca/examples/ary_13_8/ary_13_8.htm>

Six sigma
<http://www.isixsigma.com/>
<http://www.fnal.gov/pub/ferminews/ferminews01-03-16/p1.html>

Standard deviation in spreadsheets
<http://www.beyondtechnology.com/tips016.shtml>

▲▼▲

STEP FUNCTIONS

A *step function* is a mathematical relationship that has a graph that looks like steps. As a result, the function has the same output for multiple input values. For example, a telephone company may charge you 12 cents a minute for a long-distance call. A 3.3 minute, 3.7 minute, or 4.0 minute call will be charged 48 cents, or the price of a four-minute phone call, because the phone rate rounds up for every fraction of a minute beyond a whole value. In this case, the price of the phone call in dollars, p, can be determined by the function $p = 0.12\lceil t \rceil$, where t is the length of the phone call in minutes. The $\lceil t \rceil$ indicates that the value for t should be rounded up to the nearest integer. This type of step function is called a *ceiling function* and is sometimes represented by the expression $ceil(t)$. Therefore the phone-call function can also be written as $p = 0.12ceil(t)$. Any phone call between 3.01 and 4.00 minutes will result in the same charge, or any phone call between 4.01 and 5.00 minutes will result in the same charge, and so on. The figure below illustrates the price of a phone call as a function of its time.

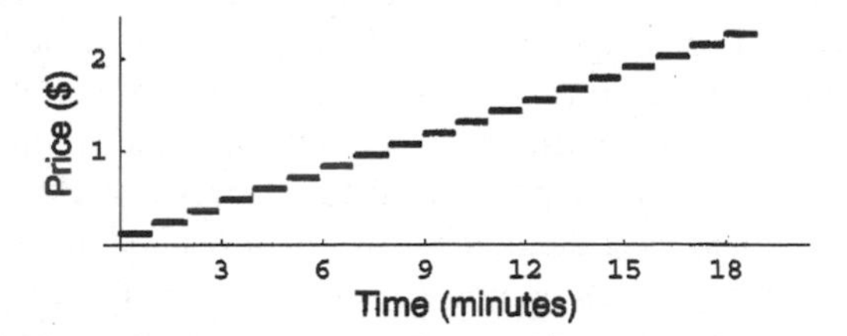

The total price of a long-distance phone call as a function of its length when the fee is 12 cents per minute or any fraction thereof.

Other rates that use discrete values for pricing can often be modeled with step functions. The price to mail a package is dependent on its mass according to a step function. If the cost to deliver a letter is 34 cents for the first ounce and 23 cents for each additional ounce, then the function $p = 0.23\lceil m - 1 \rceil + 0.34$ describes the total price in dollars, p, as a function of the mass in ounces, m. This equation is slightly different than the one for the price of a phone call, because there is a different rate for the first ounce. The $\lceil m - 1 \rceil$ portion of the equation accounts for the additional price of any mass above one ounce. You can determine this relationship in the equation because any value of m between 0 and 1 will cause the quantity $\lceil m - 1 \rceil$ to equal 0, meaning that nothing additional to 34 cents will be added to the cost of postage for mail that is between 0 and 1 ounces.

Consulting and repair rates are often represented by step functions. A visit to an attorney's office might be \$100 for making an appointment, and then an additional \$150 per hour, or fraction thereof. That means that an hour-and-a-half appointment would be equivalent to a \$400 fee—\$100 for showing up and \$300 for two hours of work. Sometimes rates are divided into smaller increments of time, such as with automobile repair. Some auto shops may charge \$80 per hour, and make charges to the next one-half hour. That means that a car that has been repaired for an hour and 13 minutes will be charged for 1.5 hours of labor, or \$120. As a step function, the repair cost in dollars, r, in terms of the number of hours of labor, h, is represented by the equation $r = 40\lceil 2h \rceil$. This equation needs to consider the number of half-hour intervals, since the overall charge is rounded to the nearest half-hour. The $2h$ in the equation describes the number of half-hours of labor, and the 40 represents the half-hour rate of \$40.

The cost of a taxicab ride also relates to a step function in terms of the distance traveled. Often there is an initial amount charged for getting in the cab, like \$2.70, and then an additional fee, like \$0.30, for every block or fraction of a block traveled. In this case, a 9.3-block cab ride would cost $2.70 + 0.30\lceil 9.3 \rceil$, or \$5.70. Notice that the distance traveled would be equivalent to 10 blocks, since there is not a specific fee for 0.3 blocks. In fact, in most cases involving fees or costs paid by the consumer, rates are usually rounded up with a ceiling function.

It is easier to charge someone for partial time or expense than to give that person an added bonus.

A case in which expense is rounded down is in the payment of hourly wages. If an employee works 40.7 hours in a week, then he or she might only get paid for 40 hours time, since she did not put in a full 41 hours. A step function is used, because it is easier to pay employees at an hourly rate than a minute rate, as well as to encourage employees to follow a tight schedule. In this situation, the step function that rounds down is called a *floor function*, or the *greatest integer function*. If the employee earns \$12 per hour, then his or her weekly salary payment in dollars, s, as a function of the number of hours worked, h, is $s = 12[h]$. The $[h]$ is the symbol to represent the greatest integer value of h, which in essence rounds the value down to the nearest integer. This equation can equivalently be written as $s = 12\lfloor h \rfloor$ or $s = 12$ floor (h) so that they include symbols describing the floor function.

A floor function has also been used to identify the day of the week for any date on the calendar since 1582. The remainder of the division in the equation

$$w = \frac{d+2m+\left[\frac{3(m+1)}{5}\right]+y+\left[\frac{y}{4}\right]-\left[\frac{y}{100}\right]+\left[\frac{y}{400}\right]+2}{7}$$

is used to predict the day of the week, w, where Sunday is the first day of the week and Saturday is the seventh or zero day. The variable d represents the day of the month, m represents the number of the month, and y represents the year. An exception to the value of m is in January and February, which are the month numbers according to the previous year. That means that January is represented by 13, February by 14, March by 3, April by 4, and so on. For example, February 16, 1918 occurred on a Tuesday, because the remainder is equal to 3 when $d = 14$, $m = 16$, and $y = 1918$ are substituted into the equation.

online sources for further exploration

Calendars
<http://astro.nmsu.edu/~lhuber/leaphist.html>
<http://www.smart.net/~mmontes/ushols.html>

Find hourly rates
<http://www.allfreelance.com/>

Houston Lighting and Power calculator
<http://www.energydotsys.com/lgscalc.htm>

Postage rate calculators
<http://postcalc.usps.gov/>
<http://wwwapps.ups.com/servlet/QCCServlet [updated 4/26/01]>
<http://www.federalexpress.com/us/rates/>

Telephone rate calculator
<http://www.geocities.com/WallStreet/5395/ratecalc.html>

SURFACE AREA

There are more uses of *surface area* than determining how much paint to buy to paint a house. The mathematics of surface area determines how objects retain heat, how cans are cut from sheets of metal, how cells exchange fluids, and how animal metabolism relates to size. Two important mathematics questions about surface area are: "What shapes make surface area a minimum for a specific volume?" and "For the same shape, how do volume and surface area change as the figure is scaled up or down?"

The first question has some simple results for common figures. The cube is the solid that minimizes surface area for a specific volume in a prism. The sphere is the solid that minimizes surface area for *any* volume. This last result shows up in soap bubbles or oil drops. In the absence of other forces, these will be spheres.

Packaging companies have additional minimization issues to handle when they determine how a package such as a cereal box or a soda can should be constructed from raw materials. The desired volume is not the only issue they must consider. If the product is going to grocery stores, then it has to have standard dimensions. The shape of the product may determine or restrict the dimensions of the package. If the carton is glued together, then additional surface is needed for the glued regions. Finally, most packaging is cut from one piece of flat material, so the engineer has to decide how the cuts will be made to minimize waste. Some of the issues have natural solutions. For example, the first illustration in the figure below shows a wasteful method of cutting circular-can lids from sheets of aluminum. The middle diagram shows that stacking the circles like the cells in a beehive would produce four more lids from the same sheet of material. The complexity of cutting single cartons is shown by a flattened box of bandages in the last illustration. Many of these cartons must be cut from large pieces of glazed cardboard.

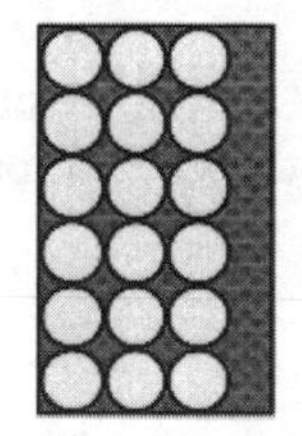

inefficient way of cutting circles (18 circles)

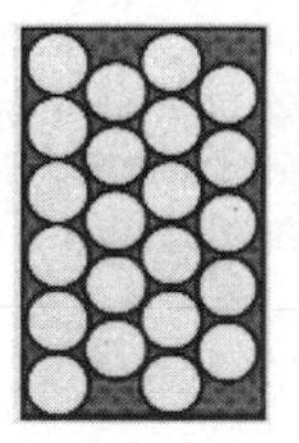

cutting circles with less waste (22 circles)

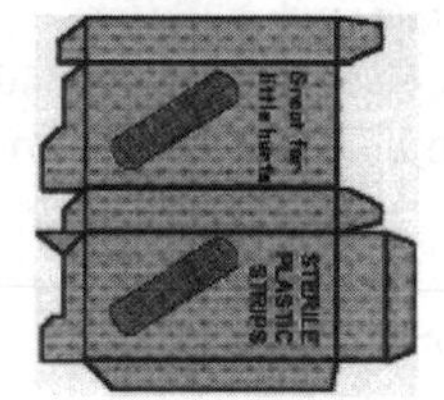

pattern for cutting a bandage box

Cutting shapes from sheets of material that will be used in packaging.

Nature has solved the minimization issue in remarkable ways. In a beehive, each cell is a regular hexagonal prism, open on one end and with a trihedral angle at the other. The trihedral angle must have a consistent geometry, because the bees build identical cells on the other side of one wall of cells. It is believed that this shape developed because it is strong and because it uses the least amount of wax (surface area) for the necessary storage area (volume of the cell). The surface area of the cell is given by $S = 6sh - \frac{3}{2}s^2\cot(\theta) + \left(\frac{3s^2\sqrt{3}}{2}\right)\csc(\theta)$, where S is the surface area, s is the length of the sides of the hexagon, and h is the height. The values of s and h are constant for specific species of bees. Using calculus, the angle that requires the least volume of wax for cells has a size of 55° regardless of s and h. Measurements of the actual angles in hives rarely differ from this value by more than 2°.

Nature sometimes needs to maximize surface area. The interiors of your lungs are networked with air sacs (alveoli). The sacs are formed from very thin membranes that allow oxygen to pass from the air in the lungs to your blood, and carbon dioxide to move from your blood to the air that will be exhaled. The surface area covered by a human's skin is about 2 square meters, but the total surface area of the alveoli is about 100 square meters! The massive surface area is needed to provide sufficient exchange of the two gasses within the time of one breath. Similarly, fish have gills that offer substantial surface membranes between the water and the bloodstream so that they can quickly exchange the carbon dioxide in blood for oxygen from the water.

Some common household tasks favor larger areas. If you want to dry wet clothes, you should spread them out rather than rolling them into a ball. If you want to cool a drink fast, crush an ice cube into the beverage rather than dropping a solid cube into it.

The *fundamental law of similarity* asserts that when you scale up (or down) a solid figure by a scale factor k, you scale up the surface area by k^2 and the volume by k^3. If you build a car model that is a 1:24 scale model of a real car, that means you are multiplying each dimension of the car by 1/24. The surface area would be changed by a factor of $(1/24)^2$, and the volume by $(1/24)^3$. If the model and the real car were made from the same materials, then the weight scale would match volume. Weight would be scaled down by $(1/24)^3$. (See ***Ratio***.) Because scaling has such a dramatic influence on surface area and volume, larger animals have an easier time maintaining their metabolism levels than do smaller ones. This can be shown by examining the ratio of volume to surface area for a series of cubes, starting with 1 cm on a side through 1 meter on a side.

side (cm)	*area (cm²)*	*volume (cm³)*	*ratio of volume to area*
1	6	1	1/6 ≈ 0.17
3	54	27	27/54 ≈ 0.5
10	600	1,000	1,000/600 ≈ 1.67
50	15,000	125,000	125,000/15,000 ≈ 8.33
80	38,400	512,000	512,000/38,400 ≈ 13.33
100	60,000	1,000,000	1,000,000/60,000 ≈ 16.67

Ratio of volume to area for different cubes.

An animal loses heat through external surface area. The energy needed for basic metabolism is roughly proportional to the volume of the animal. An animal that has a large volume with respect to its surface area will have an easier time maintaining its metabolism. A large animal will have a lower heart rate and food requirements that are a fraction of its weight. An animal that has a small volume compared with surface area will have to work hard to replace the heat lost. We can expect small animals to have rapid heartbeats and daily food requirements that may be multiples of the animals' weight. A polar bear is compact: Its large bulk means that it will have a large volume-to-area ratio. A hummingbird has a volume-to-surface-area ratio close to 0.5. It loses heat rapidly and hence must have a rapid heartbeat and relatively large food intake to maintain its metabolism. Perhaps that is one of the reasons that there are no hummingbirds in the Arctic.

Of course, animals aren't cubes. A human being is not packaged like a polar bear. We have hands that are very useful, but in winter they provide more surface area proportional to the volume they contain, so the wise person will wear mittens instead of gloves to reduce the heat loss.

The volume-to-surface-area ratio is a factor at the microscopic level. Cells cannot benefit from larger volume-to-surface-area ratios. Since energy must come through the cell membrane, small ratios are an advantage. The compensation made by plant cells is that larger plant cells maintain less of a spherical shape (more cylindrical), while small plant cells are close to spherical. With a less spherical shape, the larger cells maintain an advantageous volume-to-surface-area ratio.

online sources for further exploration

Camping (wearing gloves)
<http://www.princeton.edu/~oa/winter/wintcamp.shtml>

Figuring out how many rolls of wall covering you need
<http://www.homerepairworkshop.com/scripts/hrw.mv?ACTN=DSPLY&ART=146>

Lead paint concentrations
<http://www.cpsc.gov/cpscpub/pubs/lead/leadapp2.html>

Medicinal tablet surface area
<http://www.micromeritics.com/sa_gem_a91.html>

Minimum surface area of a can
<http://jwilson.coe.uga.edu/emt725/MinSurf/Minimum.Surface.Area.html>

Painting
<http://www.resene.co.nz/archspec/datashts/olsurf.htm>

Soap bubbles
<http://www.exploratorium.edu/ronh/bubbles/bubbles.html>
<http://micro.magnet.fsu.edu/featuredmicroscopist/deckart/index.html>

Size effects on airplane lift
<http://www.grc.nasa.gov/WWW/K-12/airplane/size.html>

Theory of flight
<http://web.mit.edu/16.00/www/aec/flight.html>

Ultracapacitors
<http://www.powercache.com/products/technical.html>

Unfolding the human brain
<http://scientium.com/drmatrix/sciences/math.htm>

SYMBOLIC LOGIC

The nineteenth-century mathematician George Boole is the acknowledged founder of modern *symbolic logic*. He recognized that an algebra of logic could be developed following the model of the algebra of real numbers. The variables of the algebra are statements that have one of two values: TRUE (1) or FALSE (0). The fundamental operations are NOT, AND, and OR, as opposed to the number operations of opposite, multiplication, and addition. The algebra of logic underlies decision-making, modern electronics, library searches, and branching in computer programs.

Computer software is built on logical structures. If a market analyst had to select all of the California residents who were female from a computer database of customers in the United States, the status of each person in the database would be evaluated with an expression such as "this person is from California and this person is a women." Sarah Jones (female) from San Francisco would evaluate as TRUE for both parts. Amy Redfox (female) from Arizona would be evaluated as FALSE AND TRUE. Since Amy misses one criterion, she should be excluded from the final set. Hence the value for Amy should be FALSE AND TRUE = FALSE. Although there may be millions of people in the database, each one falls into only one of four categories of logic. If you use a spreadsheet program, you can build *truth tables* that show the relationships among NOT, AND, and OR for these four cases. In the table below, all possible cases of statements *p* and *q* are listed using the words TRUE and FALSE. Each case has been evaluated with the spreadsheet's functions for logic. For example, the spreadsheet formula in cell D2 is = AND(A2,B2). The formula in cell E4 is = OR(A4,B4). The last column gives truth values of an expression that uses of all of the Boolean operations.

p	*q*	*NOT p*	*p AND q*	*p OR q*	*(p AND q) OR (NOT q)*
True	True	False	True	True	True
True	False	False	False	True	True
False	True	True	False	True	False
False	False	True	False	False	True

Truth tables from Microsoft Excel.

Each of these has a representation in an electrical circuit. The diagram below is a circuit showing two ON–OFF switches *p* and *q*. The circuits pass through AND, OR, and NOT connectors that act on the current as though it were a logic statement, with ON represented by TRUE, and OFF by FALSE. When will the light bulb be on? The logical expression corresponding to the circuit is in the last column of the spreadsheet in the previous table. The light is off when *p* is FALSE (OFF) and *q* is TRUE (ON). All other situations result in the light being ON.

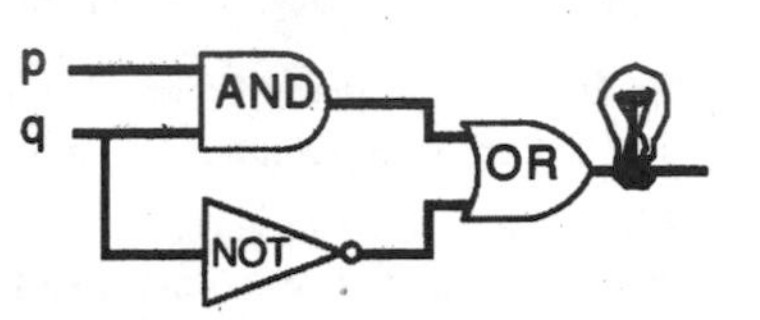

An electrical circuit using logic switches.

Two Boolean expressions that yield the same truth tables are *equivalent*. When complex circuits are expressed as Boolean algebra statements, the rules of logic can be used to simplify the circuit to one that is logically equivalent. The result is lower cost. Some circuits are used so frequently that they are designed as "new" Boolean operations. One of *DeMorgan's laws* is that (NOT *p*) OR (NOT *q*) is equivalent to NOT(*p* AND *q*). The first form would require a circuit with three logic switches. The second requires only two. The result is usually combined in a switch called a NAND switch. There is also a NOR switch that computes NOT(*p* OR *q*).

Computers represent numbers in binary form, whereby the numbers 0, 1, 2, 3, 4, 5, 6 look like 0, 1, 10, 11, 100, 101. The digits in a base 2 number can be stored as a sequence of memory positions (bits) that are on (1) or off (0). Addition rules for three cases of digit pairs are easy: $0 + 0 = 0$, $0 + 1 = 1$, $1 + 0 = 1$. The third case requires a "carry": $1 + 1 = 10$. Circuits called "half-adders" perform the addition of two bits to produce a sum bit and a carry bit. The addition of multidigit numbers requires many half-adders.

Boolean operators are the fundamental connectors in written commands that perform searches on the Internet or in computer-based library card-catalogs. Inquiries on such databases are called "Boolean searches." The set operations of union and intersection are used in place of OR and AND, respectively, in set theory.

The example of the light circuit assumes that electricity flows through a circuit instantaneously. Circuits that represent sequential firing of switches require that the algebra include a parameter for time. Although this complicates the operations, a time parameter makes the Boolean operators effective for describing neural nets in the brain and spinal cord, as well as simplifying computer circuits that require timed pulses of electricity.

Boole suggested that the truth values of 1 and 0 could be extended to probabilities of a statement being correct. In the late 1960s his idea was formalized in the field called *fuzzy logic*. The algorithms for fuzzy logic related to the binary logic shown here, but have been more successful in providing answers to problems that start with vague or contradictory information. Applications have included the computer recognition of handwritten Japanese Kanji characters, home-use blood-pressure indicators, and recognition of trends in masses of information on stock prices. Fuzzy logic procedures appear in such diverse applications as determining the length of hospital stays, detecting insurance fraud, deciding where to drill for oil, and selecting the television slot time that would be the best for an advertiser.

online sources for further exploration

The DOIT Information Science online course (select "hardware")
<http://doit.ort.org/course/intro.htm>

Boolean search vocabulary
<http://www.netstrider.com/search/logic.html>
<http://www.health.library.mcgill.ca/eguides/boolean.htm>

Logic in the computer language C
<http://www.learn-c.com/boolean.htm>

Logic in humor: Monty Python's argument clinic
<http://www.infidels.org/news/atheism/sn-python.html>

Logic in rhetoric
<http://www.sjsu.edu/depts/itl/graphics/main.html>

Fuzzy logic
<http://www.doc.ic.ac.uk/~nd/surprise_96/journal/vol1/sbaa/article1.html>

SYMMETRY

Many everyday objects are *symmetrical*. Something is reflection-symmetric if it is divided into two equal pieces so that one piece can fold directly over the other piece. The folding line is called the line of symmetry, and is also a reflecting line. Numerous types of insects are reflection-symmetric, such as butterflies and ants.

The wings of butterflies are reflection-symmetric with their bodies.

Human faces are nearly reflection-symmetric, but no one is perfect! Seldom are human feet perfectly symmetrical, since one foot is typically slightly larger than the other. However, shoes are manufactured to be symmetrical because the same foot is not larger on every person nor is the larger foot uniformly larger among people. Hence, one foot may have a tighter fit in one of the shoes.

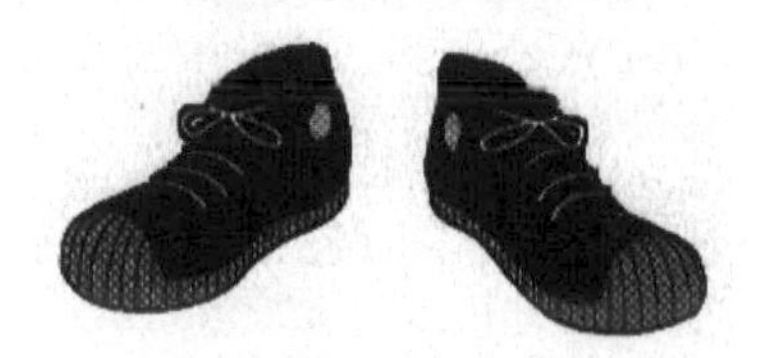

Shoes are reflection-symmetric. The size and shape of the soles of each shoe match when they are placed on top of each other.

Some objects are designed to be reflection-symmetric so that they can balance and have more support. For example, airplane engines and wheels are placed equidistant from the fuselage to divide equally their mass and power. Certain merchandise might be intentionally built not to be reflection-symmetric if it is customized to meet a person's needs. For instance, scissors are made especially for right-handed or left-handed people. A left-handed person will find it more difficult cutting with a right-handed pair of scissors.

Many kites are reflection-symmetric, because they have a cross beam that is a perpendicular bisector of the other cross beam, as shown in the following figure. The perpendicular bisector then becomes a line of symmetry that divides the kite into two equal pieces.

The cross beams in a kite divide it in equal pieces and provide support when the kite is in the air.

An object is rotation-symmetric if one of its pieces can be rotated around a point so that it is congruent with its other piece(s). For example, some flowers are rotation-symmetric, because their pedals are uniformly distributed, as shown below.

The pedals of a flower are rotated around its center and are evenly spaced apart.

Some items are constructed to be rotation-symmetric so that they can be useful at multiple angles, or provide an equal distribution. For example, wrenches can turn bolts at many different positions, and screwdrivers can twist screws continuously. An eggbeater is equally productive at all of its angles when mixing cookie batter. The blade at the bottom of the lawn mower cuts the grass evenly. A fan helps circulate air continuously and equally. Playing cards can be held either right-side up or upside down. A quarterback can throw a smooth spiral to optimize the distance of a throw because a football is shaped symmetrically.

An object that is intentionally produced so that it is *not* rotation-symmetric sometimes serves a unique purpose, such as that it can only be used in one position or does not want its mass distributed equally. For example, a knife is intended to be held by its handle; a gun can only be fired in one direction; and a pitcher of water has a handle and lip to provide more support and smoother pouring.

online sources for further exploration

Frieze patterns
<http://www.ucs.mun.ca/~mathed/Geometry/Transformations/frieze.html>

Occupations
<http://www.kitezh.com/symmetry/>

Oriental carpets
<http://mathforum.org/geometry/rugs/>

Symmetry activities
<http://www.camosun.bc.ca/~jbritton/jbsymteslk.htm>

Symmetry and the shape of space
<http://comp.uark.edu/~cgstraus/symmetry.unit/>

Symmetry around the world project
<http://www.schools.ash.org.au/stkierans-manly/Classes/Yr6/6B/Symmetry/>

Symmetry, crystals, and polyhedra
<http://www.uwgb.edu/dutchs/symmetry/symmetry.htm>

Symmetry in physics
<http://www.emmynoether.com/>

Symmetry point groups
<http://newton.ex.ac.uk/people/goss/symmetry/>

Symmetry project
<http://www.stleos.pvt.k12.ca.us/StLeosSite/classes/Seventh/realworldgeometry/SYMMETRY/symindex.html>

Types of trusses
<http://www.trussnet.com/Resources/Basics/types.cfm>

Wallpaper groups
<http://www.clarku.edu/~djoyce/wallpaper/>

▲▼▲

TANGENT

The term *tangent* can be used to describe a function (see ***Periodic Functions***) or a ratio in trigonometry applications (see ***Triangle Trigonometry***). A geometric tangent is a segment or line that locally touches a curve or figure at one point, but does not pass through the curve at that location. For example, $y = x^3 - 3x^2 + 2x - 7$ has a tangent of $y = 2x - 11$ at the point (2,–7), as shown below.

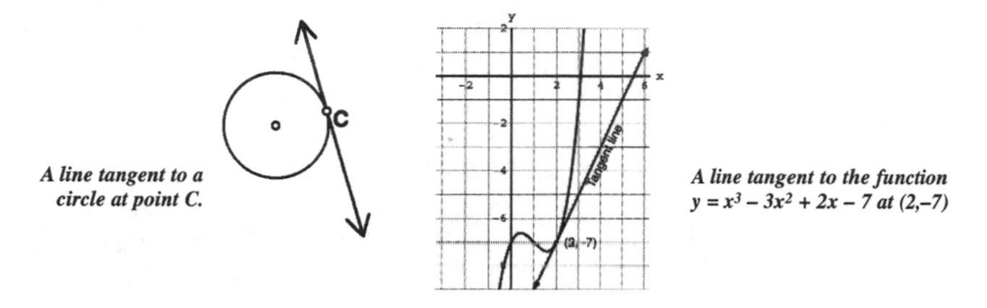

A line tangent to a circle at point C.

A line tangent to the function $y = x^3 - 3x^2 + 2x - 7$ at (2,–7)

A line can be tangent to many types of curves, including geometric shapes and functions.

The slope of a tangent line represents the derivative of a function at a point. This value is the same as the instantaneous rate of change of an object with varying rate. For example, the number of bushels, b, of corn removed in a field can be modeled with the function $b = 50 - 50e^{-0.08h}$, where h is the number of hours past 8:00 AM. The rate of productivity during any hour of the day can be determined by evaluating the derivative with a specific value of h, which is the same as the slope of the line tangent to the curve at that point, as shown in the figure below. Without the derivative, the slope of the tangent line can be approximated by finding the slope of a secant line that contains two points that are extremely close to the point of tangency. For example, $h = 4$ at 12:00 PM. The production rate at noon can be approximated by the slope of the line between 11:59:59 and 12:00:01. These times should be converted into decimals so that they can be substituted into the equation. Since there are 3,600 seconds in an hour, the difference of 1 second from 12:00 PM will be measured as 1/3600, or approximately 0.000278 hours, from $h = 4$. Using the slope formula, $m = \frac{y_2 - y_1}{x_2 - x_1}$, the slope of the tangent is

$$m \approx \frac{b(4.000278) - b(3.999722)}{4.000278 - 3.999722} \approx \frac{.001615}{.000556} \approx 2.9 \text{ bushels per hour}.$$

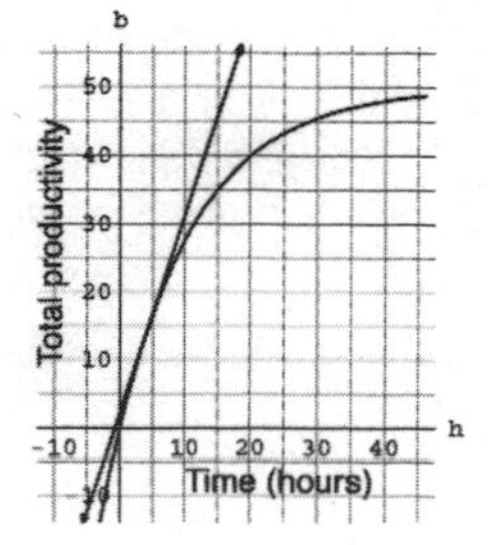

The slope of the tangent line can predict the productivity rate at a particular moment in time.

At this rate, it would seem appropriate to take a break so that workers can rejuvenate for the afternoon.

Besides worker productivity, the slope of a tangent line can help determine the speed of an object, the location where business profits are at a maximum, the hourly rate for business consulting, the moment when ticket sales for a particular movie have declined rapidly, and many other applications about rates that can be modeled with functions (see ***Rates***).

Tangents are also used in applications related to circles. For example, radio signals will reach a distance from the antenna on the tower to the horizon. The visible sight to the horizon represents the point of tangency, where no other parts of the earth can be seen. Since a line tangent to a circle is perpendicular to its radius, this distance can be determined using the Pythagorean theorem. A radio antenna that is 200 meters tall can have a signal that reaches a distance of approximately 50 km. Since the radius of the earth is approximately 6,380 km, the equation $s^2 + 6,380^2 = 6,380.2^2$ is used to find the signal radius, s, based on the geometric representation depicted below.

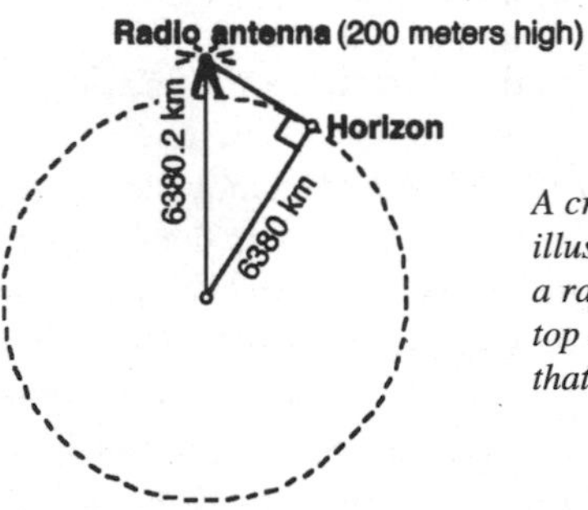

A cross-sectional view of the earth that illustrates the broadcasting distance of a radio antenna—the distance from the top of the tower to the horizon. Note that the diagram is not drawn to scale.

An object released from a circular-motion path will leave in a straight-line tangent to the circle at the point of release. For example, a discus thrower spinning in a circle will throw the disc out towards the open field in a straight path after she releases the weight from her hand, as shown below. When a cowboy spins a lasso in a circular path, and then releases his grip, the rope will travel in a straight path towards the calf that he is trying to capture.

A discus thrower rotates rapidly to add momentum to a throw. Even though the thrower is rotating as she releases the discus, the projectile will be along a straight path.

The smoothness in sidewalk curves is designed using common tangents from arcs on different circles. Without the use of tangents, curved sidewalk paths would have jagged corners, as depicted as follows.

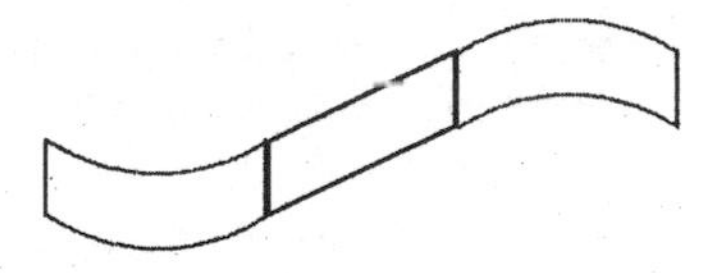

sidewalk curves with nontangent arcs of circles

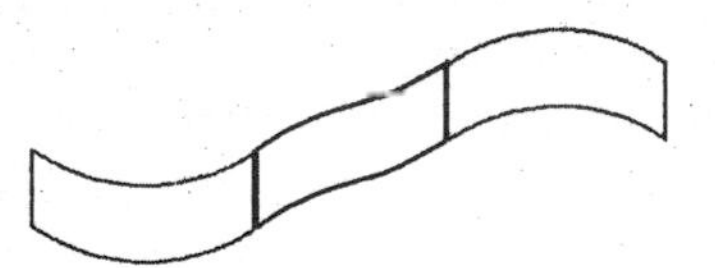

sidewalk curves with tangent arcs of circles

Smooth curves in sidewalks are created with tangent arcs of circles.

Belts that operate machinery, as shown in the figure below, are wrapped around circular wheels that keep the belts in motion as they rotate. Since the belts are tangent to both circles, they can smoothly cycle around the wheels without jumping or falling off.

Moving belts remain tightly on spinning wheels, because they are tangent to both circles at both locations.

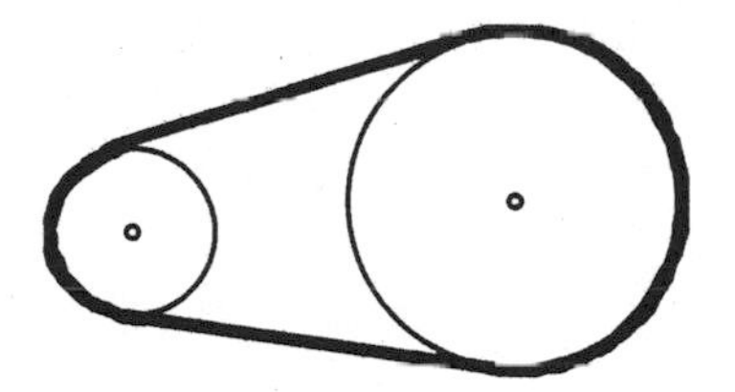

online sources for further exploration

Centripetal force
<http://www.glenbrook.k12.il.us/gbssci/phys/Class/circles/u6l1c.html>
<http://regentsprep.org/Regents/physics/phys06/bcentrif/default.htm>

Curved mirrors
<http://www.math.montana.edu/frankw/ccp/calculus/deriv/mirror/learn.htm>

Eliminating the discharge snub pulley
<http://www.mineconveyor.com/snubhead.htm>

Ferroelectric complex oxides
<http://www.sas.upenn.edu/chem/gallery/phys/rappe.html>

Introducing the ellipse (flashlight)
<http://www.geocities.com/CapeCanaveral/Lab/3550/ellipse.htm>

Selection and installation of conveyer belt scales
<http://www.rocktoroad.com/selection.html>

Surfing
<http://www.ies.co.jp/math/java/calc/doukan/doukan.html>

▲▼▲

TRANSLATIONS

A *translation* is a shift of points over the same distance and in the same direction. When you slide a checker piece across a game board from one square to another, you are performing a translation. A home run in baseball represents the hitter's four translations: home plate to first base, first to second, second to third, and third to home. The carpet design that is produced in hundreds of yards of a carpet roll represents many translations of a single design. Musicians who transpose a piece of music down to the range of a singer are performing a translation.

Translations in the coordinate plane can be expressed by the addition of coordinates. The following figure shows the translation of the plane by the translation 6 right and 2 down. The translation can be expressed as an ordered pair (6,–2), and the transformation by addition of ordered pairs. This is shown on the drawing as the movement of a triangle. Every point (x, y) in the preimage of the triangle will be translated 6 right and 2 down to a corresponding point (x', y') in the image triangle. This gives the equation $(x, y) + (6,\text{-}2) = (x', y')$. Applying this to the vertex $(\text{-}3, 4)$ gives $(\text{-}3, 4) + (6,\text{-}2) = (3, 2)$ as shown. Applied to the vertex $(\text{-}5, 1)$ gives $(\text{-}5, 1) + (6,\text{-}2) = (1,\text{-}1)$. The picture shows that the corresponding vertex in the image triangle is $(\text{-}1, 1)$.

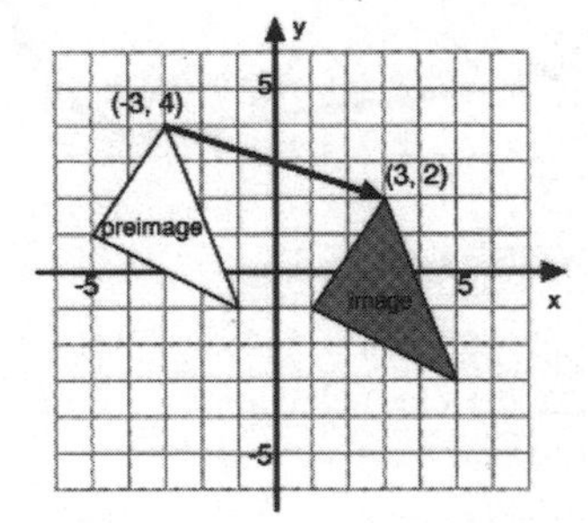

The translation (slide) of a triangle by 6 right and 2 down, $(x, y) + (6,\text{-}2) = (x', y')$.

Translations change equations of functions in a systematic fashion. If a graph of a function $y = f(x)$ is translated by (h, k), the resulting graph will be the function $y - k = f(x - h)$. If you translate a parabola $y = x^2$ by (3,-2), the resulting graph has the equation $y - (\text{-}2) = (x - 3)2$, or $y = x^2 - 6x + 7$. Translations apply to three-dimensional figures and functions in the same way as they do in two dimensions.

The language of translations depends on the application. Translations that represent moves of chess pieces can be indicated by the column or rank (labeled with letters in the picture below) and rows or files (labeled with numbers).

Coordinate system in chess.

Coordinate notation indicates the starting square for the piece and the ending square. The white knight can move B1-C3. This represents the translation, or move, of the knight. Chess players keep track of games and communicate with distant players using this coordinate system or the similar algebraic system.

When you sing “Frere Jacques” or “Row, Row, Row My Boat” with other people, it is likely that you separate into groups. When the first group finishes the line “Row, row, row my boat, gently down the stream,” the second group will start singing. When it finishes the first line, the third group will start. Meanwhile, the first two groups continue singing. Songs that are melodious when the start is shifted by line are called *rounds*. The shift is a translation in time.

Shifts in musical keys are called *transpositions*. Shown below is a four-note theme from a Mozart symphony in C major transposed down to G major. Each note has been shifted six piano notes down.

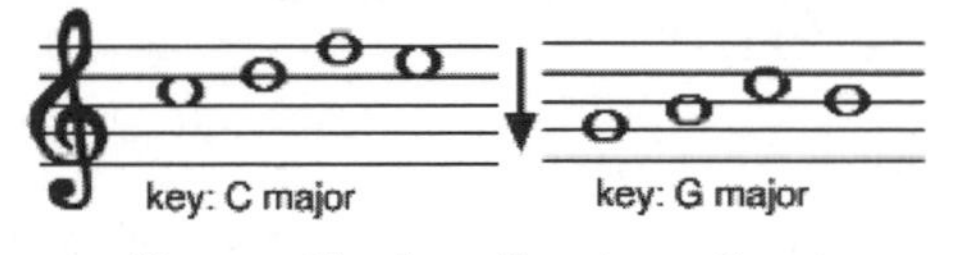

Transposition from C major to G major.

Translations have been used to build patterns in painting, architecture, weaving, and ceramics from ancient Greeks to medieval Celts to contemporary Acoma Pueblo potters in New Mexico. The basic ribbon pattern is based on repeated translations of a simple design (see the left figure below). Another design is based on translating the basic figure, then reflecting it. This is called a *glide reflection* (right figure, below). Translations, rotations, and reflections of “seed” patterns are fundamental in designing quilt patterns.

Left: A translation pattern from Pueblo pottery. Right: A glide reflection—a translation alternated with a reflection.

The simple translation of a simple seed figure, as in the left illustration, is the basis of periodic functions. A seed, such as the two nodes of the sine function from 0° to 360° can be translated in 360° moves to create the full periodic function of the sine. (See ***Periodic Functions***.)

online sources for further exploration

Art

<http://hometown.aol.com/Cyrion7/celtic/index.htm>

Music

<http://www.musictheory.halifax.ns.ca/20key_trans.html>

TRIANGLE TRIGONOMETRY

Trigonometry can be used to find unknown lengths or angle measurements. In a situation involving right triangles, only a side length and an angle measurement are needed to determine the length of an object. This information is useful to engineers, because they can find large or hard-to-measure distances without having to measure them. For example, the height of a flagpole or a tall building can be determined using a measured distance from the pole and an angle of elevation from the ground (see below).

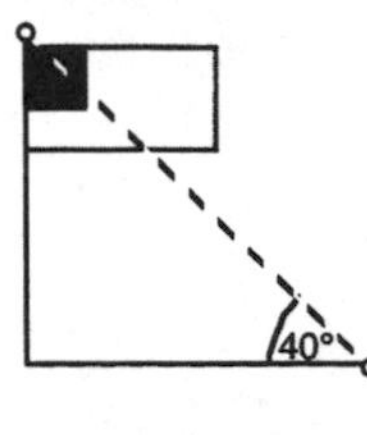

The height of a tall object, such as a flagpole, can be determined with trigonometry by measuring a distance along the ground and an angle of elevation.

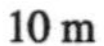

Suppose you measure a distance 10 meters away from a flagpole along the ground. You record an angle of elevation at that point equal to 40°, as depicted. In right-triangle trigonometry, one of the following three ratios can be used to find the flag height of the poll:

$$\sin\theta = \frac{\text{opposite}}{\text{hypotenuse}} \qquad \cos\theta = \frac{\text{adjacent}}{\text{hypotenuse}} \qquad \tan\theta = \frac{\text{opposite}}{\text{adjacent}}.$$

In this case, $\tan\theta$ (pronounced *tangent*) should be used, because the opposite side from the angle of elevation θ is unknown (the height of the flagpole), and the adjacent side is the distance along the ground of 10 meters. Therefore the height of the flagpole, approximately 8.4 meters, can be found by solving the equation $\tan 40° = \frac{\text{flagpole}}{10}$.

Engineers use transits to measure angles of tall or hard-to-reach objects.

Sometimes the angle of elevation is recorded from an object above the ground, such as a transit sitting on a tripod, as illustrated. If the angle measurement is not taken from the ground, then the height of the tripod will need to be included in the final calculation. In this case, if the transit is 1.5 meters off of the ground, then the angle of elevation would be approximately 35° (see the following figure).

The missing length will be approximately 6.9 meters, after setting up an equation using the tangent function, $\tan 35° = \frac{\text{length above transit}}{10}$. To find the complete flagpole length, the height of the transit will need to be added to this calculation in order to obtain the same answer calculated earlier.

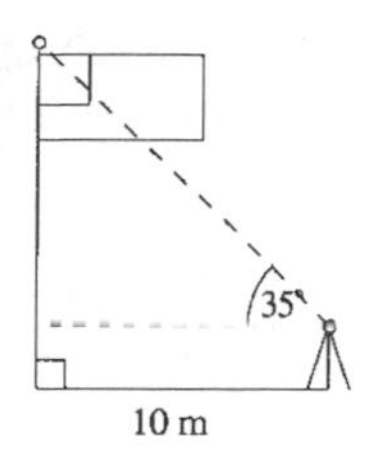

A transit can be used to measure the angle of elevation to help determine the height of a tall object, such as a flagpole. The sum of the height of the transit and the leg of the right triangle along the flagpole represents the total height of the flagpole.

Right-triangle trigonometry can be used to determine an unknown angle based on two lengths. For example, the navigator of a ship will try to minimize the traveling distance by adjusting the direction of the boat to account for the water's current. If the current is moving parallel to the waterfront, then the speed of the boat observed from land will be greater due to the push from the current. Suppose that the ship is moving perpendicular to the shore at 40 feet per second and is recording a land speed of 42 feet per second.

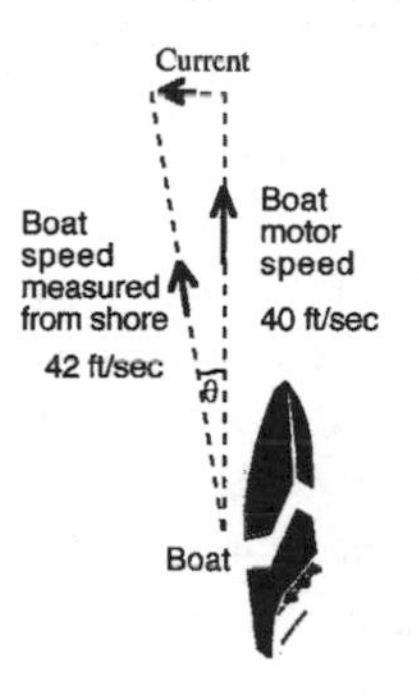

Boats need to turn an angle by θ against the current in order to account for the force of it so that they can head in the most direct path towards the shore.

The current will push the boat off course if it is trying to reach a destination directly across the river. Using the cosine of the angle $\cos\theta$, the ship's navigator can determine the angle in which to rotate the boat so that it does not move off course. The cosine function is used in this case, because the two measurements known are the *adjacent* (the boat speed) and *hypotenuse* (the land speed) sides of the right triangle. Substituting the given values in this relationship, the unknown angle of 17.8° can be found by solving the equation $\cos\theta = \frac{40}{42}$. To find an angle measurement, the inverse cosine of the ratio, or $\cos^{-1}(40/42)$, needs to be entered on the calculator. This means that if the boat moves straight towards its journey, it will actually veer off course by 17.8°. If the boat is still headed straight without accounting for the current, it will veer almost one-third of a mile off course for every mile traveled. To avoid this problem, the ship's navigator will have to turn the boat 17.8° away from the perpendicular path and against the current in order to travel directly across the river.

Applications of right-triangle trigonometry also exist in areas outside of surveying and navigation. Air-traffic control at small airports must establish the cloud height in the evening to determine if there is enough visibility for pilots to safely land their planes. A light source directed at a constant angle of 70° towards the clouds situated 1,000 feet from an observer, and the observer's angle of elevation θ to the spotlight in the clouds, are sufficient information to determine the cloud height (see below).

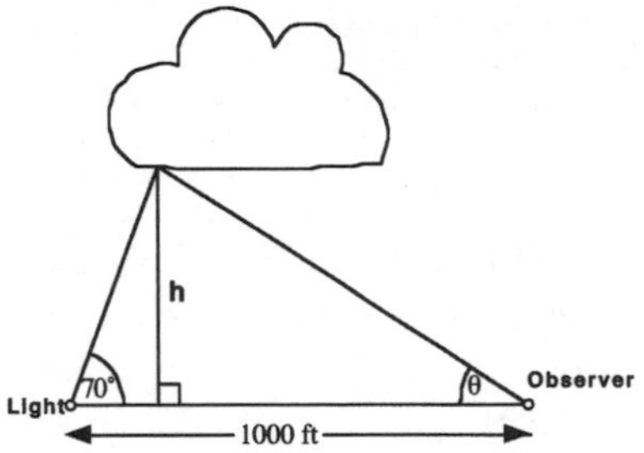

Determining the cloud height.

In this situation, the equations $\tan\theta = \frac{h}{y}$, $\tan 70° = \frac{h}{x}$, and $x + y = 1,000$ can be used to find the cloud height, h. Planes can safely land if the cloud height is above 1,000 feet, with horizontal ground visibility of at least three miles.

The pilot can also use right-triangle trigonometry to determine the moment when a plane needs to descend towards the airport. If the plane descends at a large angle, the passengers may feel uneasy due to a quick drop in altitude and also may not adjust well to changes in pressure. Consequently, the pilot tries to anticipate the opportunity to descend towards the airport at a small angle, probably less than 5°. Based on the plane's altitude, air-traffic control at the airport can determine the point at which the plane should begin to descend. With a descent angle of 3° and altitude a, the plane should start its approach at a distance of $\frac{\tan 3°}{a}$ feet away from the airport, assuming that the plane descends at the same angle until it reaches the ground.

Construction workers can determine the length of a wheelchair ramp based on restrictions for its angle of elevation. For example, suppose an office needs to install a ramp that is inclined at most 5° from the ground. If the incline is too great, it would be difficult for handicapped people to move up the ramp on their own. Based on this information, the architect and construction workers can determine the number of turns needed in the ramp so that it will fit on the property and stay within the angle-of-elevation regulations. In addition to wheelchair ramps, a similar equation can be set up to determine the angle by which to pave a driveway so that an automobile does not scrape its bumper on the curb upon entering and leaving.

All triangle applications finding unknown sides or angles, however, are not always situated in settings where a right triangle is used. In these cases, either the *law of sines* or *law of cosines* can be applied. One example of applying the law of sines is to find the height of a hill or a mountain, since it is unlikely that one will be able to find the distance from the base of a hill or mountain to its center, as shown in the following figure.

The law of sines states that the ratio of the sine of an angle to the side length of its opposite side is proportional for all opposite angle and side pairs. That is, in triangle ABC, $\frac{\sin A}{a} = \frac{\sin B}{b} = \frac{\sin C}{c}$. If a person measures an angle of eleva-

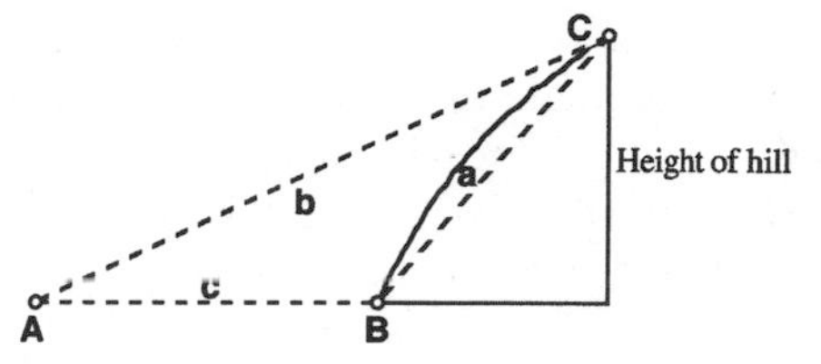

The height of a hill can be determined using the law of sines and right-triangle trigonometry by measuring the angles of elevation at points A and B, and the distance between points A and B.

tion from the base of the hill to its peak, and then repeats the measurement at a given distance away, the law of sines can be used to find the height of the hill. Actually, it can first be used to find the length along the side of the hill, and then right-triangle trigonometry can be used to find the hill's height. In this case, a surveyor takes measurements $c = 1,000$ feet apart and measures angles of elevation to the tip of the hill equal to $m < B = 75°$ and $m < A = 43°$. The following equation to find the length alongside the hill, a, can be set up using the law of sines: $\frac{\sin 32°}{1{,}000} = \frac{\sin 43°}{a}$.

The 32° angle opposite the 1,000 foot distance can be found by using the fact that the sum of the angles in a triangle is equal to 180°. This length of a, approximately 1,287 feet, can help engineers determine the amount of railway needed to build a funicular to transport materials, or the amount of cable needed to build a gondola line for skiing. Since a right triangle is in the diagram, right-triangle trigonometry can be used to find the hill's height. Solve the equation $\sin 75° = \frac{h}{1{,}287}$ to determine the height of the hill, h, which is approximately 1,243 feet. That is a length equal to about four football fields, but straight up in the air!

The law of cosines is a theorem used in triangle trigonometry to find the measurement of a side when two sides and an included angle are given, or to find the measurement of an angle when three sides are given. For example, a public-works contractor can determine the amount of cement needed to pave a new road that intersects two other intersecting roads in town (to form a triangle), as shown below.

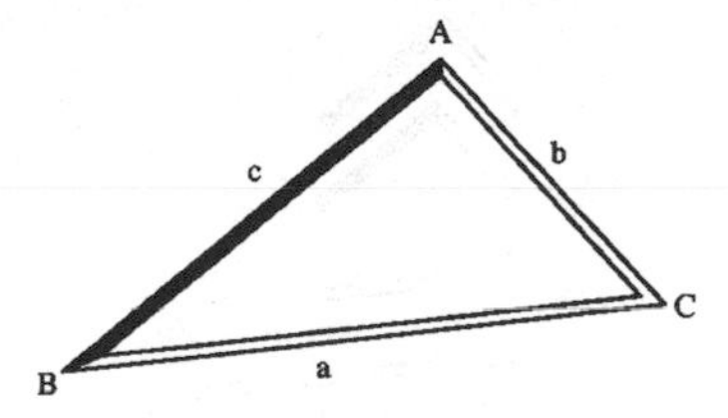

The length of a newly paved road can be determined using the law of cosines, given the length of two adjacent roads and angle C—the angle between the existing road.

In this case, the contractor needs to determine the angle formed between the existing roads, $m\angle C$, and the location of the intersection of the other two roads in order to predict the distance of the new road. Since the distance traveled is proportional to the amount of cement used, the formula $c^2 = a^2 + b^2 - 2ab \cos C$ will help determine the amount of cement needed to connect the roads, where a, b, and c are sides of the triangle (the length of the roads), and C is the angle included between the existing roads a and b. A similar type of investigation would also be needed for bridge designers or tunnel developers.

Triangle trigonometry has many other applications that help find unknown lengths or angle measurements. For instance, paintings, motion pictures, and televisions have ideal viewing distances in order to create the greatest possible image from the eye. The triangle is formed between the view and the top and bottom (or the sides) of the viewing object.

Astronomers use triangle trigonometry to determine distances and sizes of objects. For example, the distance from the earth to the moon, and earth to the sun, can be found by identifying their angles from the horizon during an eclipse. The height of a solar flare can also be determined by measuring the angle from the sun to the tip of the flare, and using distance information about the earth and sun.

online sources for further exploration

Astronomy and geography
<http://www.geocities.com/Hollywood/Academy/8245/trigonometry.html>

Cable television
<http://www.wake.tec.nc.us/math/Projects/Raychem/deb-raychem_trig.htm>

Civil engineering and navigation
<http://www.ece.utexas.edu/projects/k12-fall98/14545/Group1/app.html>

Height of a tree
<http://jwilson.coe.uga.edu/emt725/Kite/kite.html>
<http://www.math.bcit.ca/examples/ary_15_3/ary_15_3.htm>

Inclined ramp
<http://www.math.bcit.ca/examples/ary_12_3/ary_12_3.htm>

Pipe flow
<http://www.math.bcit.ca/examples/ary_8_3/ary_8_3.htm>

Saturn's mass and distance from earth
<http://www.amtsgym-sdbg.dk/as/AOL-SAT/SATURN.HTM>

Surveying
<http://www.math.bcit.ca/examples/ary_17_3/ary_17_3.htm>
<http://catcode.com/trig/trig13.html>

▲▼▲

VARIATION

When two quantities increase proportionally, we say they are *directly* related. Mathematically, the two quantities x and y must be related as $y = kx$, where k is a nonzero constant called the *constant of variation*. The formula for circumference c of a circle in terms of radius r is $c = 2\pi r$. The constant of variation is 2π. The independent variable can be a power. The area of a circle is directly related to the *square* of the radius, $A = \pi r^2$. The constant of variation is π. Kepler's third law of planetary motion uses a fractional power. The period T of a planet's orbit around the sun is proportional to the 3/2 power of its distance R from the sun, $T = kR^{3/2}$. Because Kepler's first law stated that planets circle the sun in elliptical paths, the semimajor axis provides the measure of distance.

Many geometry formulas can be expressed as direct variation. Since the area of a cube is $A = 6s^2$, where A is the surface area and s is the length of an edge, it follows that $s = \sqrt{\frac{A}{6}}$. The length of an edge varies directly as the square root of surface area of the cube. The length of the edge varies directly as the cube root of the volume V, $s = \sqrt{V^3}$.

Joint variation occurs when the dependent variable varies directly as the product of two or more independent variables. Many geometry formulas are in joint variation. The volume of a cylinder is $V = \frac{1}{3}\pi r^2 h$. The volume V varies jointly as the radius r squared and the height h. The constant of variation is $\frac{1}{3}\pi$. The volume of a rectangular solid having length L, width W, and height H is expressed in the formula, $V = LWH$. The volume varies jointly as length, width, and height. The constant of variation is 1.

Biologists and medical scientists have provided formulas for the surface area of a human-being's skin. The DuBois formula relates area in square centimeters jointly to the 0.425 power of weight in kilograms and the 0.725 power of height in centimeters, $A = 71.84W^{0.425}H^{0.725}$. The formula estimates the surface area for the average adult male to be about 1.8 square meters, and for the average adult female, about 1.6 square meters.

Population biologists use different kinds of variation to express rates of change. The change in a population undergoing rapid growth (see ***Exponential Growth***) is $c = rP$, where c is the change in the number of organisms, P is the population count before change, and r is the rate of change. In 1995, Mexico's population was 91.1 million people. It was increasing at a rate of 2.0 percent per year. The change formula for Mexico would be the direct variation formula, $c = 0.02P$. Using the formula to predict the change in population for 1995 to 1996 gives, $c = 1.822$ million people. The change for the following year would be based on 92.2 people. If there is a limit to the population of a country, say M people, then the change formula would be $c = kP(M - P)$. Change in a population varies jointly as the current population and the available capacity for people. This leads to a more complex pattern of growth. (See ***Logistic Functions***.)

Inverse variation occurs when the variables are related through a reciprocal. If you must travel 200 miles at a constant rate, the distance-rate-time formula says that $200 = rt$. Solving for t gives the equation $t = \frac{200}{r}$. In this equation, t varies inversely as r. The constant of variation is 200. The independent variable can be a power. For example, the intensity I of light falling on an object varies inversely as the *square* of the distance d from the light. The formula is $I = \frac{k}{d^2}$. (See ***Inverse Square Function***.)

The law of the lever is an inverse variation. The distance d from the fulcrum in feet needed to stabilize the seesaw with a person who weights w pounds is

$d = \frac{k}{w}$. If Jane weighs 100 pounds and sits 5 feet from the fulcrum, how far away will Juan, who weighs 150 pounds, have to sit in order to balance Jane? Use Jane's data to find the constant of variation k: $5 = k/100$, so $k = 500$. Now solve for Juan's distance: $d = 500/150 = 3.33$. Juan would have to sit 3 feet 4 inches from the fulcrum in order to balance Jane. Note that the constant k was computed from Jane's statistics. If she were to change position or be replaced by someone else, the value of k would change.

Pulley systems are a series of ropes and wheels that help lift and support heavy objects by distributing weight in multiple locations. Elevator shafts rely on pulleys to move the cabin, and movers use pulleys to transport cumbersome or heavy objects such as pianos into tall buildings. A 100 pound weight can feel like a 50 pound weight when it is moved by a two-pulley system, because half the weight is distributed at the other pulleys. As the number of pulleys in the system increases, the amount of force needed to move the object decreases proportionally. Therefore a three-pulley system needs a 33.33-pound force to move the 100 pound weight, a four-pulley system needs a 25 pound force to move the 100 pound weight, and so on. The force, f, needed to move an object, the weight of the object, w, the number of pulleys needed in a system, p, are related with the equation, $f = w/p$. If the weight is constant, then the force applied varies inversely with the number of pulleys used.

Compound variation combines direct and indirect variation with two or more independent variables. The gravitational force between two planets varies directly as the product of the masses of the planets, and inversely as the square of the distance between them: $F = \frac{Gm_1m_2}{d^2}$, where F is the force in newtons, G is a gravitational constant (6.67×10^{-11} newton-meters per square kilogram), r is the distance in meters between the centers of two planets, and m_1 and m_2 are the mass of each planet in kilograms. The constant of variation would be different if measurements are made in different units, such as in feet rather than meters and pounds rather than kilograms. The formula works if one of the planets is the earth and the other "planet" is a person high above the earth's surface. It simplifies to an inverse-variation formula for the weight of a body above the earth: $W = \frac{k}{d^2}$, where W is the weight above the planet, d is the distance between the person and the *center* of the earth, and k is a constant. It may seem strange that both masses have disappeared, but they are handled by the constant. Consider a 170 pound astronaut who is 9,000 miles above the surface of the earth. How much does he weigh at that altitude? First write the equation for his weight at the surface of the earth. Since the radius of the earth is about 4,000 miles, $170 = \frac{k}{4{,}000^2}$. Solving for k yields, $k = 2.72 \times 10^9$. The inverse-square formula is therefore $W = \frac{2{,}720{,}000{,}000}{d^2}$. Using this formula with the distance $d =$ 13,000 miles from the center of the earth gives, $W \approx 16.09$. The astronaut would weigh about 16 pounds.

The deflection D of a diving board is a function of the weight W of the diver, the length of the board L, the elasticity E of the material making up the board, and the moment of inertia I of the cross section of the board. The variation formula is $D = k\frac{WL^3}{3EI}$.

Ohm's law is a direct variation statement $V = IR$, where V is voltage, I is current, and R is the resistance in a particular conductor. R, which is measured in ohms, is constant of variation for the particular conductor. Resistance is measured in ohms and will vary across different wires. For example, electrical resistance R in a wire varies directly as its length L, and inversely as its cross-sectional area A: $R = \frac{\rho L}{A}$, where ρ is the constant of variation. The constant of variation is called *resistivity* and has been computed for many materials: gold has a resistivity of 2.35×10^{-8}; carbon, 3.50×10^{-5}; and wood, 10^8. If one assumes that the wire is round, then the variation is $R = \frac{kL}{r^2}$. The coefficient of variation, k, would be the resistivity divided by π. If resistance in a wire must be reduced, there are two routes: you could shorten the wire, or you could use a wire with a larger radius. The latter might have the most payoff, because the radius is squared in the formula. The three-dimensional graph below shows the resistance (vertical axis) of copper wire wrapped into a coil. The lower-left axis shows the radius in meters of wire running from 2 mm up to 1 cm (0.01 meter). The axis on the right shows how long the wire would be if it were unwrapped. The axis runs, right to left, from 0 to 1,200 meters. The length does not appear to affect results. However, radii under 5 mm send the resistance soaring.

Resistance of coiled copper wire for wire diameters from 0.002 to 0.01 m, and lengths from 0 to 1200 m.

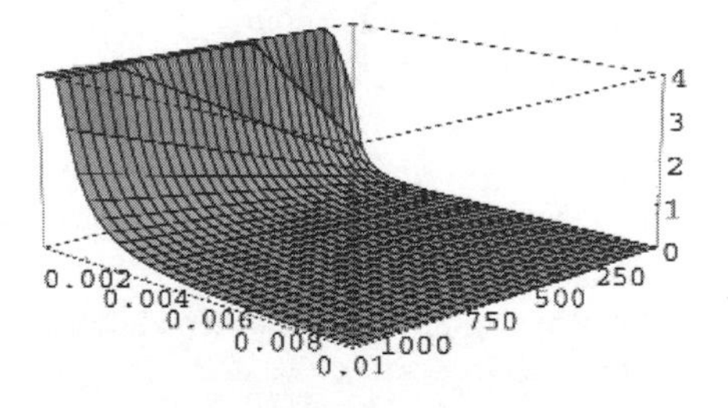

Some laws appear in different forms of variation depending on the situation. The simple form of Poiseuille's law states that the speed S at which blood moves through arteries and veins varies directly with the blood pressure P and the fourth power of the radius r of the blood vessel: $S = kPr^4$. This is derived from Poiseuille's law for the flow of fluids, which relates to flow F rather than speed (flow is speed times cross-section area of the tube): $F = \frac{k\Delta Pr^4}{nL}$, where ΔP is the change in pressure from the beginning of a tube to the end, r is the radius of the tube, n is a measure of viscosity of the fluid, L is the length of the tube, and k is a constant of variation. In this version, which is used to determine the flow of oil through pipes and also fluids through tubes in an automobile, flow is directly related to the change in pressure and fourth power of tube radius, and is inversely related to viscosity and the length of tube.

In general, solving variation problems involves two steps: first, solve for the constant of variation in a known situation; and second, use that constant to rewrite the formula and evaluate the dependent value for the unknown situation. Variation problems can also be solved with proportions. (See ***Proportions***.)

online sources for further exploration

Applications of variation
<http://www.iln.net/html_p/c/72782/62079/53795/53836/58708_58712.asp>
<http://www.jcoffman.com/Algebra2/ch9_2.htm>

Diving
<http://library.thinkquest.org/28170/34.html>

Fan laws
<http://www.apco1650.demon.co.uk/fdr.htm>

Financial hedging
<http://www.cs.trinity.edu/~rjensen/000overview/mp3/138intro.htm>

Galilieo's pendulum experiments
<http://es.rice.edu/ES/humsoc/Galileo/Student_Work/Experiment95/galileo_pendulum.html>

Kepler's laws
<http://www.cvc.org/science/kepler.htm>
<http://observe.ivv.nasa.gov/nasa/education/reference/orbits/orbit3.html>

Harmonics, resonance, and interference
<http://www.sasked.gov.sk.ca/docs/physics/u5c42phy.html>

Murphy's law of locksmithing
<http://www.google.com/url?sa=U&start=7&q=http://www.jfbdtp.com/Murphy.html&e=747>

Population growth described in Annenberg Math in Everyday Life
<http://www.learner.org/exhibits/dailymath/population.html>

Ventilation
<http://human.physiol.arizona.edu/SCHED/Respiration/Morgan31/Morgan.L31.html>

Formulas that show different variation can be found in XREF
<http://www.xrefer.com/>

VECTORS

Vectors emerged from the study of physical situations in which two or more forces were applied to an object. A vector is a directed line segment whose length is proportional to the time, distance, or force being measured. Many people use vectors to give directions. "Go down this street six blocks. Turn left at the stop sign. Go another two blocks to the stoplight. Turn left at the stoplight at Maple

Street. Go southwest eight blocks on Maple." Applications of vectors appear in physics, navigation, computer graphics, airplane design, and statistics.

In the illustration below, Sarah and James are pulling a wagon. Sarah pulls with 25 pounds of force, and James with 15 pounds. They are pulling at an angle of 20°. What is the net direction and force on the wagon? The James vector $\vec{j}$ is drawn along the x-axis of the grid. The three units represent 15 pounds of force. The vector for Sarah, $\vec{s}$, is 5 units long, representing 25 pounds of force. It is drawn 20° from $\vec{j}$. The parallelogram law in physics states that the resultant vector $\vec{r}$ is the diagonal of the parallelogram that is formed with sides parallel to the vectors. (See ***Quadrilaterals***.) The obtuse angle in the triangle is formed by two sides and the diagonal is 160°. By the law of cosines (see ***Triangle Trigonometry***),

$$r^2 = s^2 + j^2 - 2sj\cos\theta$$
$$r^2 = 25^2 + 15^2 - 2(25)(15)\cos 160°.$$

Solving for r yields a force of 39.43 pounds. Using the law of sines computes the angle between $\vec{r}$ and $\vec{j}$ to be about 12.5°. Because they are pulling at an angle, the forces don't add to the total possible for Sarah and James (40 pounds of force), but they come close.

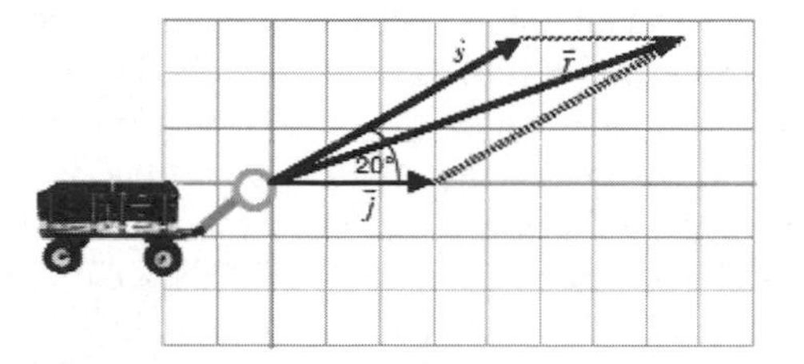

Vectors showing James $\vec{j}$ and Sarah $\vec{s}$ pulling on a wagon. Their resultant force is $\vec{r}$, indicated by the diagonal of the parallelogram.

The same kind of analysis applies to paths of airplanes. The next figure shows an airplane pointed due northeast at 400 miles per hour. Its vector $\vec{a}$ is drawn 45° clockwise from north. A 90-mile-per-hour wind is blowing 10° south of east. The wind vector $\vec{w}$ is shown at 10° clockwise from east. The angle between the plane vector and the wind vector is 55°. The plane will be blown somewhat off course. What is its direction and ground speed? The situation depicted indicates directions in degrees according to navigation conventions. Complete the resultant vector $\vec{r}$ and compute its length and direction. From the

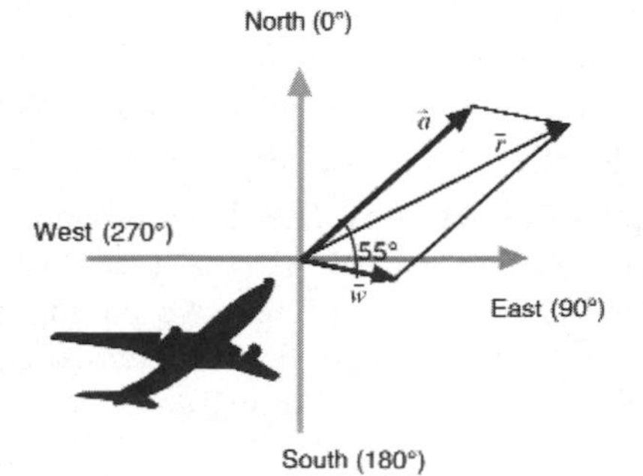

Vector diagram for an airplane headed northeast that is blown off course by a wind 10° south of east.

law of cosines $r^2 = 400^2 + 90^2 - 2(400)(90)\cos 125°$. The length of $\vec{r}$ is about 458 miles per hour. The angle between $\vec{a}$ and $\vec{r}$ is about 9.3°. So the direction would be about $45° - 9.3° = 35.3°$ north of east. Even though the airplane would be pointed northeast, from the ground it would appear to be traveling only 35.3° north of east. (See ***Triangle Trigonometry***.)

When vectors are written as an ordered pair, the length is written first, and the angle second. (See ***Polar Coordinates***.) Sarah's vector would be written as $\vec{s} = [25, 20°]$; James' vector would be $\vec{j} = [15, 0°]$. The brackets indicate that the vector is written in polar-coordinate form. The lengths of the vectors are written with the absolute value sign. The length of Sarah's vector would be $|\vec{s}| = 25$. Polar form is a natural way of presenting force vectors, but the algebra of vectors is easier to work with in Cartesian-coordinate form (*x*,*y*). This is called the *component* form. To convert a vector in polar form $\vec{v} = [d, \theta]$ to component form, use the formulas $x = d\cos\theta$ and $y = d\sin\theta$. Sarah's polar vector would be $\vec{s} = (25\cos 20°, 25\sin 20°) \approx (23.50, 8.55)$. To reconstruct the length of Sarah's vector from component form, use the Pythagorean theorem:

$$|\vec{s}| = \sqrt{(25\cos 20°)^2 + (25\sin 20°)^2} = 25.$$

The addition of vectors in component form is done by the addition of coordinates. If $\vec{v} = (a, b)$ and $\vec{w} = (c, d)$, the parallelogram law requires that the vector sum be $\vec{v} + \vec{w} = (a + c, b + d)$. Component form makes it easier to handle problems involving gravity. If a golf ball is hit with an impact of 70 meters/second at a 30° angle, the distance of the ball (ignoring wind resistance and gravity) is given by the vector $\vec{b} = [70t, 30°]$, where time t is given in seconds. The component form is $\vec{b} = (70t\cos 30°, 70t\sin 30°)$. A graph would show the golf ball traveling upwards into space at an angle of 30° from the ground. However, gravity provides a force vector that reduces vertical distance as $g = (0, -4.9t^2)$. The vector addition of the ball and gravity gives a parabolic path produced by $\vec{b} + \vec{g} = (70t\cos 30°, 70t\sin 30° - 4.9t^2)$. Algebra can be used to determine how far the ball has traveled horizontally when it hits the ground. (See ***Angle*** for computations of the path of a projectile.) Vector descriptions of motions and forces are used to describe the collisions of atomic particles, the interaction of chemical substances, and the movements of stars and galaxies.

Component form has operations that are somewhat like multiplication, but yet different. The *dot product* of two vectors is given by $\vec{v} \bullet \vec{w} = (ac, bd)$, where $\vec{v} = (a, b)$ and $\vec{w} = (c, d)$. Lengthening a vector by a scale factor k is given by $k\vec{v} = (ka, kb)$. The dot product is used in the formula for the cosine of an angle between two vectors: $\cos\theta = \frac{\vec{v}\bullet\vec{w}}{|\vec{v}||\vec{w}|}$. The effectiveness of component-form vectors comes when vectors operate in more dimensions. For three-dimensional space, the dot product of $\vec{v} = (v_1, v_2, v_3)$ and $\vec{w} = (w_1, w_2, w_3)$ is $\vec{v} \bullet \vec{w} = (v_1w_1, v_2w_2, v_3w_3)$, an easy-to-remember extension of the two-component model. Further, the equation for the cosine of the angle between two vectors looks exactly the same, even though there is an additional dimension.

Since the concepts of addition, dot product, and angle between vectors scale up to many dimensions, vector mathematics adapts well to statistical computations. Consider the test-score data on five students shown in the table below. The deviation scores form vectors with five components. The science vector is $\vec{s} = (5, 0, 1, -1, -5)$. The math vector is $\vec{m} = (5, -3, 3, -2, -3)$.

student	*science raw score* X	*math raw score* Y	*science deviation score* x	*math deviation score* y
Albert	85	25	5	5
Manuel	80	17	0	-3
Bonnie	81	23	1	3
Sharon	79	18	-1	-2
Elena	75	17	-5	-3
average (mean)	**80**	**20**		

Test data on five students. Deviation scores are computed by taking the test score minus the mean (e.g., $x_{\text{Bonnie}} = 81 - 80 = 1$;. $y_{\text{Elena}} = 17 - 20 = -3$).

Computations with the vectors give the lengths to be

$$|\vec{s}| = \sqrt{5^2 + 0^2 + 1^2 + (-1)^2 + (-5)^2} = \sqrt{52} \text{ and } |\vec{m}| = \sqrt{56}.$$

When each of these lengths is scaled by the reciprocal of the square root of dimensions, $\frac{1}{\sqrt{5}}$, the computation produces the *standard deviation* for each vector. These are about 3.22 for $\vec{s}$ and 3.34 for $\vec{m}$. (See ***Standard Deviation*** for uses and formulas.) The cosine of the angle between the two vectors is

$$\cos\theta = \frac{\vec{s} \bullet \vec{m}}{|\vec{s}||\vec{m}|} = \frac{5(5)+0(-3)+1(3)+(-1)(-2)+(-5)(-3)}{\sqrt{52}\sqrt{56}} \approx 0.83.$$

This is called the *correlation coefficient* for the two vectors and is commonly designated with the letter r. We say that for this group of students, science scores correlate 0.83 with math scores. Because it is a cosine, the correlation coefficient r ranges from -1 to +1. Correlations at +1 (angle $\theta = 0°$) and -1 (angle $\theta = 180°$) indicate that the vectors are collinear. Correlations close to 0 (angle $\theta = 90°$) indicate that the vectors are going in different directions. In the first case ($r = 1$), the vectors are pulling in the same direction. In the second case ($r = -1$), they are opposites. Our correlation coefficient for science and math tests ($r = 0.83$) corresponds to an angle between the vectors of about 33.5°. In a space of five dimensions, these vectors are separate enough that each one is measuring some underlying skills that are different for different students, but they are also measuring something that is the same for all students. Generally, students who scored high on science also scored high on math. The square of the cosine provides a measure of overlap. This *coefficient of determination* is $r^2 = 0.83^2 \approx 0.70$. It indicates that 70 percent of the variability in math scores is accounted for by the variability in science scores. When statisticians work with many scores, they examine the correlations among all the variables to determine how the number of dimensions of the original space can be reduced to fewer, stronger, and more interpretable dimensions.

In the case of three dimensions, the operation of *cross product* provides a way to compute perpendiculars to planes. The cross product of $\vec{v} = (v_1, v_2, v_3)$ and $\vec{w} = (w_1, w_2, w_3)$ is defined as $\vec{v} \times \vec{w} = (v_2w_3 - v_3w_2, v_3w_1 - v_1w_3, v_1w_2 - v_2w_1)$. The cross product is a vector. Its relationship to the plane formed by $\vec{v}$ and $\vec{w}$ is shown in the figure. The cross product is said to be *orthogonal* to the plane.

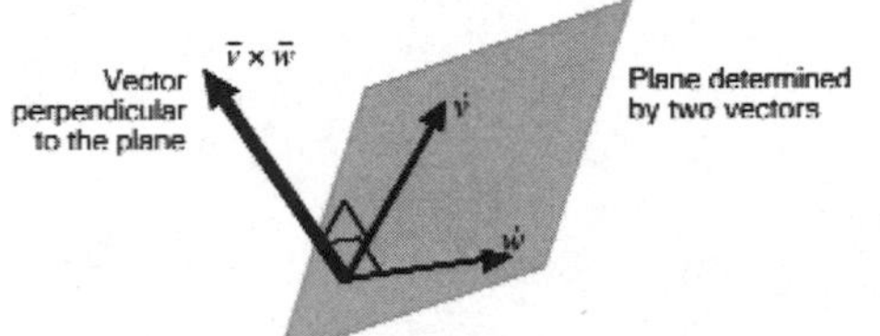

The cross-product vector $\vec{v} \times \vec{w}$ is perpendicular to the plane formed by the vectors $\vec{v}$ and $\vec{w}$.

The cross product is computed for surfaces of airplanes or boat hulls. The direction of air or water currents across the surfaces is modeled by the angles that the currents make with vectors that are orthogonal to the surface. This is not a recent concept. Sketches in the notebooks of the Wright brothers one hundred years ago show computation of vector forces on the different wings they tried before achieving the first airplane flight. A spinning wheel, like the disk in a gyroscope, produces a force called *torque*. This is a force that is perpendicular to the plane of rotation. If you ride a bike very fast, you will feel resistance as you try to tilt the bicycle to the left or right. The torque produced by the spinning wheels will try to maintain its direction, so you must use some pressure to produce a tilt. If you are traveling slowly, the torque isn't very strong, so it is easy to tilt the bike and fall. Large cruise ships have gyroscopes with heavy wheels that spin rapidly. The torque produces a force that counters the movement from waves, making for a smoother ride for passengers.

Computer-graphic programmers use orthogonal vectors to determine how light sources would hit surfaces visible in a computer game or architectural image. The angles between the light rays from an external source to orthogonal vectors on the surface are computed. If the angles are close to 0°, then the light will be shown at full intensity. If close to 90°, then the light is reaching the surface with minimal intensity. The vector computations (the *vector-graphic* phase) are then transferred into the display device as light intensity and color for the different points (pixels) that would be visible. This is the *raster graphic* phase. By controlling the brilliance of pixels on the screen according to vector computations, computer-graphic designers present realistic scenes to the viewer. Some computer files store images as vectors (the rules that create the image), and some files keep the bitmap of the image (a snapshot of the pixel intensity). Postscript files contain rules for generating graphics on printers and computer screens, and use vector concepts in drawing letters as well as pictures. Graphics files on your computer that end in .GIF or .JPG are raster files. Vector files (geometric files) are easier to modify than raster files. Raster files display faster than vector files, although the high speed of modern processors makes this a negligible difference to the ordinary computer user. Vector descriptions of images are used for computer identification of faces, for translating handwriting into computer text, for descriptions of protein structures, and for the location of tumors in medical CAT scans.

online sources for further exploration

Animations of vector operations
<http://www.reed.edu/~obonfim/Phys100.html>
<http://id.mind.net/~zona/mstm/physics/mechanics/vectors/components/vectorComponents.html>
<http://www.frontiernet.net/~imaging/vector_calculator.html>

Computer graphics
<http://www.enginemonitoring.org/illum/illum.html>
<http://www.sli.unimelb.edu.au/envis/hidden.html>
<http://www.ati.com/na/pages/resource_centre/dev_rel/sdk/RadeonSDK/Html/Tutorials/RadeonBumpMap.html>
<http://www.greuer.de/ecalc3d.html>

Magnetic Resonance Imaging (medical visualization)
<http://www.cis.rit.edu/htbooks/mri/>

Parachute vectors
<http://www.explorescience.com/>

Vector properties
<http://www.netcomuk.co.uk/~jenolive/homevec.html>
<http://forum.swarthmore.edu/~klotz/Vectors/vectors.html>
<http://www.glenbrook.k12.il.us/gbssci/phys/Class/vectors/u3l1a.html>

Vectors in text
<http://www.wdvl.com/Authoring/Graphics/Tools/PSP7/Text_Path/>

▲▼▲

VOLUME

Volume describes the amount of space contained in a three-dimensional object. Almost every object we use has volume, due to having depth. Even a sheet of paper has volume, because it has thickness, although a very thin one. If an object, such as paper, does not have volume, then it could be stacked indefinitely without having any height. You can estimate the thickness of a sheet of paper by measuring the height of a ream of paper and then dividing by 500 sheets. Hence, to find the volume of the sheet of paper, or the amount of wood needed to make the paper, you would divide the volume of the prism formed by the ream by the number of sheets of paper in the ream.

Manufactures think about volume as they build containers for their products. Canned and boxed foods are often sold by their mass. Knowing the density of the substance can help determine the amount of volume it will use in a container, since density, d, is the ratio of mass, m, and volume, v. In terms of an equation, $d = \frac{m}{v}$. Nonuniform products that contain air pockets such as potato chips and cereal will often have additional empty volume when a package is opened, because the contents will have settled and filled air pockets.

In addition to packaging food, companies that produce fragile items need to consider the volume of additional materials that are needed, such as Styrofoam, shredded paper, or packing bubbles. The amount of insulated packaging needed would be equal to the difference between the volume of the box and the volume of the item. If the item being shipped is in the form of a geometric solid, such as a prism, pyramid, sphere, or cylinder, then the volume can be predicted with an equation. For example, suppose a crystal ball with a radius of 2 inches is shipped in a cubical container with an edge length of 6 inches. The volume of packaging material needed to surround the crystal ball would be: the volume of the cube minus the volume of the sphere $= 6^3 - \frac{4}{3}\pi \bullet 2^3 \approx 182$ cubic inches. That is almost 85 percent of the space in the box!

Beverage production and distillation centers use the concept of volume to determine how many containers can be filled based on their available raw materials. Cola companies need large tanks, usually cylindrical, to mix the raw ingredients needed to create soft drinks. Once created, the cola will need to be emptied into cans for distribution. Suppose a 5,000 gallon tank of cola is ready to be dispersed into 12 ounce cans. If each gallon is equivalent to 128 fluid ounces, then 5,000 $\times$ 128 = 128,000 ounces of cola are available to produce a little more than 53,000 soft drinks (128,000/12 $\approx$ 53,333), and over 2,200 (53,333/24 $\approx$ 2,222) cases for distribution.

Ice cream cones are constructed so that the ice cream drips inside of the cone as it melts. When ice cream is served, the spherical scoops lie on top of a cone that is empty inside. The volume of ice cream inside the cone will gradually increase as the temperature of the ice cream rises and pressure is applied at the top of the cone. The cone keeps the ice cream inside it from melting more quickly, since it is not exposed to the outside air temperature. An ice cream cone with a height of 8 cm and base radius of 2 cm can hold close to half of a scoop of ice cream with radius 2.5 cm. This is determined by dividing the volume of the cone $\frac{1}{3}\pi \bullet 2^2 \bullet 8$ by the volume of the spherical scoop $\frac{4}{3}\pi \bullet 2.5^3$, whose ratio is approximately 0.512.

Construction workers who use concrete consider the amount of cement needed to complete a job. When a driveway for a new house is planned, its dimensions need to be measured so that an appropriate amount of cement is brought to the site. Suppose a driveway 12 feet wide and 30 feet long is needed for a new home, and the cement poured needs to be 1.25 feet deep so that it will not break apart under pressure from automobiles or extreme temperatures. The volume of cement needed to complete this job can be determined by the product of its dimensions, since the cement will fill into a rectangular prism. Therefore the amount of cement needed is (12)(30)(1.25) = 450 cubic feet. If a bag of cement mixture prepares 0.75 cubic feet of concrete, then the cement truck will need to contain 600 bags of mixture in order to create the driveway. Since each bag is about a hundred pounds, multiple trucks will be needed to carry the 60,000 pound load.

The dispersion of an oil spill can be predicted based on the amount of oil that is lost. On March 24, 1989, the oil tanker *Exxon Valdez* struck Bligh Reef in Prince William Sound, Alaska, spilling more than 11 million gallons of oil. There are 231 cubic inches in a gallon, so the spill had an approximate volume of 254 billion ($231 \times 11{,}000{,}000 = 254{,}100{,}000{,}000$) cubic inches of oil. Thousands of marine animals and fish were killed by the oil that contaminated the water. As oil spreads, it typically leaves a layer that is 1/100 inch thick on the surface of the water. The direction of the spill is influenced by the placement of the spill and the direction of the ocean's current. In this circumstance, much of the oil had brushed on shore, at the beaches. What if the spill had happened in the middle of the ocean? Imagine the effect of spilling water in an open space on the floor. Assuming that the floor is flat, the spill will disperse in nearly a circular region. If the oil had spilled in the middle of the ocean without land interference, the spill could have covered nearly 2.5 billion ($254{,}100{,}000{,}000 \star 1/100$) square inches of the surface of the water. If the path of the oil dispersed in the shape of a circle, then it could have spread in a radius of close to half a mile. This information can be determined by solving for r in the equation $\pi r^2 = 2{,}541{,}000{,}000$, and then converting the inch units to miles. There are 5,280 feet in one mile, and 12 inches in a foot. Therefore the conversion is $5{,}280 \times 12 = 63{,}360$ inches for every mile. Benjamin Franklin was one of the first to determine that very little oil will spread out over a huge area of water. His work actually gave one of the first estimates of the thickness of a molecule of oil, even though no one in Franklin's time knew about molecules.

online sources for further exploration

Aluminum tanks
<http://www.fifthd.com/gear/tankspecs.html>

Application to environmental health
<http://www.math.bcit.ca/examples/ary_8_1/ary_8_1.htm>

Balloon volumes
<http://www.overflite.com/volume.html>

Blood pressure
<http://www.shodor.org/master/biomed/physio/cardioweb/application.html>

Cost for landscaping
<http://gardening.sierrahome.com/tools/landscaping/volumeandcost_calc.jsp>

Density of water, ice, and snow
<http://astro.uchicago.edu/cara/southpole.edu/flaky.html>

How big are your lungs?
<http://www.troy.k12.ny.us/schools/ths/ths_biology/labs_online/school_labs/print_versions/lung_lab_school_print.html>

Measurement microphones
<http://www.josephson.com/tn6.txt>

Spherical polytropes
<http://www.phys.lsu.edu/students/valencic/approject1.html>

Tank volume
<http://www.grapl.com/vmlnotes/examples/tank_volume.htm>

Unit converter
<http://www.webcom.com/legacysy/convert2/volume.html>
<http://www.ex.ac.uk/cimt/dictunit/ccvol.htm>

Volume of an irregular solid
<http://jwilson.coe.uga.edu/emt725/Envir/Volume.html>

Bibliography

▲▼▲

Anton, H., and C. Rorres. (1987). *Elementary Linear Algebra with Applications*. New York: Wiley.

Austin, J. D. (Ed.). (1991). *Applications of Secondary School Mathematics*. Reston, VA: National Council of Teachers of Mathematics.

Bennett, J. O., W. L. Briggs, and C. A. Morrow. (1998). *Quantitative Reasoning: Mathematics for Citizens in the 21st Century*. Reading, MA: Addison-Wesley.

Boole, G. (1958). *An Investigation of the Laws of Thought*. New York: Dover.

Braun, M. (1983). *Differential Equations and Their Applications*. New York: Springer-Verlag.

Breiteig, T., I. Huntley, and G. Kaiser-Messmer. (Eds.). (1993). *Teaching and Learning Mathematics in Context*. Chichester, UK: Ellis Horwood.

Burkhardt, H. (1981). *The Real World and Mathematics*. Glasgow: Shell Centre for Mathematics Education.

Bushaw, D., M. Bell, H. O. Pollak, M. Thompson, and Z. Usiskin. (1980). *A Sourcebook of Applications of School Mathematics*. Reston, VA: National Council of Teachers of Mathematics.

Crawford, M., and M. Witte. (1999). Strategies for mathematics: Teaching in context. *Educational Leadership* 57(3): 34–38.

Evans, J. (1999). Building bridges: Reflections on the problem of learning in mathematics. *Educational Studies in Mathematics* 39: 23–44.

Gasson, P. C. (1983). *Geometry of Spatial Forms*. Chichester, UK: Ellis Horwood.

Glazer, E. M. (2001). *Using Internet Primary Sources to Teach Critical Thinking Skills in Mathematics*. Westport, CT: Greenwood Press.

Goldstein, L. J., D. C. Lay, and D. I. Schneider. (1990). *Calculus and Its Applications*, 5th ed. Englewood Cliffs, NJ: Prentice-Hall.

Gustafson, R. D., and P. D. Frisk. (1992). *Algebra for College Students*. Pacific Grove, CA: Brooks/Cole Publishing.

Hoppensteadt, F. C. (1982). *Mathematical Methods of Population Biology*. Cambridge, UK: Cambridge University Press.

Horton, H. L. (1990). *Mathematics at Work*, 3rd ed. New York: Industrial Press.
Jacobs, H. R. (1970). *Mathematics: A Human Endeavor*. San Francisco: W. H. Freeman.
Johnson, A. F. (1991). The t in $I = Prt$. In *Applications of Secondary School Mathematics*, ed. J. D. Austin (pp. 122–28). Reston, VA: National Council of Teachers of Mathematics.
Lowe, I. (1988). *Mathematics at Work: Modelling Your World*. Canberra: Australian Academy of Science.
Masingila, J. O. (1993). Learning mathematics practice in out-of-school situations. *For the Learning of Mathematics* 13(2): 18–22.
Masingila, J. O., S. Davidenko, and E. Prus-Wisniowska. (1996). Mathematics learning and practice in and out of school: A framework for connecting these experiences. *Educational Studies in Mathematics* 31: 175–200.
Mathematical Sciences Education Board, National Research Council. (1998). *High School Mathematics at Work: Essays and Examples for the Education of All Students*. Washington, DC: National Academy Press.
McCliment, E. R. (1984). *Physics*. San Diego: Harcourt Brace Jovanovich.
McConnell, J. W., S. Brown, Z. Usiskin, et al. (1996). *UCSMP Algebra*. Glenview, IL: Scott, Foresman.
Miller, C. D., and V. E. Heeren. (1986). *Mathematical Ideas*, 5th ed. Glenview, IL: Scott, Foresman.
National Council of Teachers of Mathematics. (2000). *Principles and Standards for School Mathematics*. Reston, VA: National Council of Teachers of Mathematics.
Nave, C. R., and B. C. Nave. (1985). *Physics for the Health Sciences*, 3rd ed. St. Louis: W. B. Saunders.
Nunes, T., A. D. Schliemann, and D. W. Carraher. (1993). *Street Mathematics and School Mathematics*. Cambridge, UK: Cambridge University Press.
Olinick, M. (1978). *An Introduction to Mathematical Models in the Social and Life Sciences*. Reading, MA: Addison-Wesley.
Pauling, L., and R. Hayward. (1964). *The Architecture of Molecules*. San Francisco: W. H. Freeman.
Peck, R., C. Olsen, and J. Devore. (2001). *Introduction to Statistics and Data Analysis*. Pacific Grove, CA: Duxbury.
Peressini, A. L., J. W. McConnell, Z. Usiskin, et al. (1998). *UCSMP Precalculus and Discrete Mathematics*, 2nd ed. Glenview, IL: Scott, Foresman/Addison-Wesley.
Phagan, R. J. (1992). *Applied Mathematics*. South Holland, IL: Goodheart-Wilcox.
Pozzi, S., R. Noss, and C. Hoyles. (1998). Tools in practice, mathematics in use. *Educational Studies in Mathematics* 36: 105–22.
Reagan, J. (1991). Get the message? Cryptographs, mathematics, and computers. In *Applications of Secondary School Mathematics*, ed. J. D. Austin (pp. 189–93). Reston, VA: National Council of Teachers of Mathematics.
Rhoad, R., G. Milauskas, and R. Whipple. (1984). *Geometry for Enjoyment and Challenge*, rev. ed. Evanston, IL: McDougal, Littell.
Senk, S. L., D. R. Thompson, S. S. Viktora, et al. (1996). *UCSMP Advanced Algebra*, 2nd ed. Glenview, IL: Scott, Foresman.
Smith, K. J. (1998). *The Nature of Mathematics*. Pacific Grove, CA: Brooks/Cole Publishing.
Stewart, J. (2001). *Calculus Concepts and Contexts*, 2nd ed. Pacific Grove, CA: Brooks/Cole Publishing.

U.S. Census Bureau. (1999). *Statistical Abstract of the United States: 119th Edition.* Washington, DC: U.S. Census Bureau.

Usiskin, Z., C. H. Feldman, S. Davis, et al. (1995). *UCSMP Transition Mathematics*, 2nd ed. Glenview, IL: Scott, Foresman.

Usiskin, Z., D. Hirschhorn, A. Coxford, et al. (1998). *UCSMP Geometry*, 2nd ed. Glenview, IL: Scott, Foresman.

Watson, A. (Ed.). (1998). *Situated Cognition and the Learning of Mathematics*. York, UK: QED.

Yates, D. S., D. S. Moore, and G. P. McCabe. (1999). *The Practice of Statistics*. New York: W. H. Freeman.

Zitzewitz, P. W., M. Davids, and R. F. Neff. (1992). *Physics Principles and Problem*s. New York: Glencoe.

About the Authors

EVAN M. GLAZER is a Ph.D. candidate at the University of Georgia in the Department of Instructional Technology, and a former mathematics teacher at Glenbrook South High School in Glenview, Illinois. He has written in the mathematics and technology areas. His book, *Using Internet Primary Sources to Teach Critical Thinking Skills in Mathematics*, was published in 2001 by Greenwood Press.

JOHN W. McCONNELL is a lecturer at North Park University, earning his doctorate in mathematics education from Northwestern University. Before his retirement, he was instructional supervisor of mathematics at Glenbrook South High School in Glenview, Illinois, for twenty-seven years. He has written many publications in the field of K–12 mathematics, including three textbooks in the University of Chicago School Mathematics Project.